Book I

LIFE SCIENCE ACTIVITIES FOR GRADES 2–8

Science Curriculum Activities Library

Marvin N. Tolman
James O. Morton

illustrated by Carolyn Quinton

Parker Publishing Company, Inc.
West Nyack, New York

10 9 8 7 6 5

Library of Congress Catalog Card Number: 86-61396

ISBN 0-13-536061-7

Printed in the United States of America

ABOUT THE AUTHORS

DR. MARVIN N. TOLMAN

Trained as an educator at Utah State University, Marvin N. Tolman began his career as a teaching principal in rural southeastern Utah. The next eleven years were spent teaching grades one through six in schools of San Juan and Utah counties and earning graduate degrees.

Currently associate professor of elementary education, Dr. Tolman has been teaching graduate and undergraduate classes at Brigham Young University since 1975. Subject areas of his courses include math methods, science methods, and computer literacy for teachers. He has served as a consultant to school districts, has taught workshops in many parts of the United States, and has published several articles in professional journals. Dr. Tolman is one of two authors of *What Research Says to the Teacher: The Computer and Education*, published in 1984 by the National Education Association, and a coauthor of *Computers in Education*, published by Prentice-Hall, 1986.

Dr. Tolman now lives in Spanish Fork, Utah, with his wife, Judy, and their five children.

DR. JAMES O. MORTON

For more than thirty years, James O. Morton, Ed.D., Teachers College, Columbia University, has taught students of all ages. He began his career as an elementary school teacher in the Salt Lake City, Utah, public schools, where he served later as a principal and as curriculum director. For one year Dr. Morton was a visiting lecturer in education at Queens College, New York City. He also served as an instructor on the summer faculty of Teachers College, Columbia University. For fifteen years Dr. Morton was an associate professor of science education at the University of Utah.

Dr. Morton has served as a national consultant in the development of science curriculum from early childhood through university graduate programs in all geographical areas of the United States. His publications have appeared in scientific and professional journals.

Dr. Morton currently works as a writer and science consultant and lives with his wife, Lornel, in Klamath Falls, Oregon.

ABOUT THE *LIBRARY*

The *Science Curriculum Activities Library* provides elementary teachers with over 475 science activities that give students hands-on experience in various areas. To be used in conjunction with your regular science texts, the *Library* includes three books, each providing activities exploring a different field:

- Life Sciences
- Physical Sciences
- Earth Sciences

In most public schools today, emphasis is on the fundamental skills of language arts and mathematics. The teaching of science has often been relegated to a supplementary place in the curriculum. What may be overlooked is that a strong science program with a discovery/inquiry approach can enrich the development of mathematics, as well as other academic content areas. The activities in the *Library* develop these skills. Most activities call for verbal responses, with questions that encourage analyzing, synthesizing, and inferring instead of answering yes or no.

Development of thinking and reasoning skills and understanding content are the main goals of the *Library*'s science activities. Learning how to learn and how to apply the various tools of learning is more useful in a person's life than is the acquisition of large numbers of scientific facts. Through these process skills, students are encouraged to explore, invent, and create. The learning of scientific facts is a byproduct of this effort, and increased insight and retention of facts learned are virtually assured.

HOW TO USE BOOK I: LIFE SCIENCES

Book I consists of over 140 easy-to-use, hands-on activities in the following areas of life sciences:

- Plants and Seeds
- Animals
- Growing and Changing: Animal Life Cycles
- Animal Adaptation
- Body Structure
- The Five Senses
- Health and Nutrition

Teacher Qualifications

You need not be a scientist to conduct an effective and exciting science program at the elementary level. Interest, creativity, enthusiasm, and willingness to get involved and try something new will go a long way. Two of the most critical qualities of the elementary teacher as a scientist are (1) commitment to helping students acquire learning skills and (2) recognition of the value of science and its implications for the acquisition of such skills.

Capitalize on Interest

It is expected that some areas will be of greater interest to both you and your students, so these interests should be considered when you select science topics. As these materials are both nongraded and nonsequential, areas of greatest interest and need can be emphasized. As you gain experience with using the activities, your skill in guiding students toward appropriate discoveries and insights will increase.

Organizing for an Activity-centered Approach

Trends of the past have encouraged teachers to modify a traditional textbook approach by using an activity-based program, supplemented by the use of textbooks and other materials. We favor this approach, so the following activities encourage hands-on discovery. Valuable learning skills are developed through this direct experience.

One of the advantages of this approach is the elimination of a need for every student to have the same book at the same time, freeing a substantial portion of the textbook money for purchasing a variety of materials and references, including other textbooks, trade books, audio and video tapes, models, and other visuals. References should be acquired that lend themselves developmentally to a variety of approaches, subject matter emphases, and levels of reading difficulty.

Starter Ideas

Section 1, "Starter Ideas," should be used first. The sequence of other sections may be adjusted according to interest, availability of materials, time of year, or other factors. Some sections use concepts developed in other parts of the book. When this occurs, the activities are cross-referenced so concepts can be drawn from other sections as needed.

Starter Ideas are placed at the beginning of the book to achieve several specific goals:

- To assist in selecting topics for study.
- To provide a wide variety of interesting and exciting hands-on activities from many areas of science. As students investigate these Starter Ideas, they should be motivated to try additional activities in related sections of the book.
- To introduce teachers and students to the discovery/inquiry approach. When you have experienced several of these activities, they will spend time on the other sections more efficiently.
- To be used for those occasions when only a short period of time is available and a high-interest independent activity is needed.

Unique Features

The following points should be kept in mind while using this book:

1. It places the student in the center of the discovery/inquiry approach to hands-on learning.
2. The main goals are problem solving and the development of critical-thinking skills. Content is a spinoff, but is possibly learned with greater insight than if it were the main objective.
3. It attempts to prepare teachers for inquiry-based instruction and to sharpen their guidance and questioning techniques.
4. Most materials recommended for use are readily available in the school or at home.
5. Activities are intended to be open and flexible and to encourage the extension of skills through the use of as many outside resources as possible: (a) the use of parents, aides, and resource people of all kinds is recommended throughout; (b) the library, media center, and other school resources, as well as a classroom reading center related to the area of study, are essential in teaching most of the sections; and (c) educational television programs and videocassettes can often enrich the science program.
6. With the exception of the activities labeled "teacher demonstration" or "total-group activity," students are encouraged to work individually, in pairs, or in small groups. The teacher gathers and organizes the materials, arranges the learning setting, and serves as a resource person. In many instances, the materials listed on an index card with the procedure are all students will need in order to perform the activities. Ideas are given in "To the Teacher" at the beginning of each section and in

"Teacher Information" at the end of each activity to help you develop your content background and your questioning and guidance skills.

7. Full-page activity sheets, such as "Animal Study Sheet" and "My Daily Exercise Chart," are offered throughout this book. These sheets can easily be reproduced and kept on hand.

At the end of the book are a bibliography, sources of free and inexpensive materials, and a list of science supply houses. This information will help you organize your program so that each activity is used to its fullest potential.

Format of Activities

Each activity in this book includes the following information:

- *Activity Number*: Activities are numbered sequentially throughout the book for easy reference.
- *Activity Title*: The title of each activity is in the form of a question that can be answered by completing the activity and that requires more than a simple yes or no answer.
- *Special Instruction*: Some activities are intended to be used as teacher demonstrations or whole-group activities, or require close supervision for safety reasons, so these special instructions are noted. Some activities involving plants and animals are recommended for certain seasons of the year, so be sure to read "To the Teacher" at the beginning of the sections dealing with plants and animals before you plan your science program for the year.
- *Materials*: Each activity lists the materials needed. The materials used are easily acquired, but, when necessary, special instructions or sources have been given. In some sections involving animals, inexpensive and easily obtainable live specimens are recommended. Some may need to be ordered or collected well in advance, so see "To the Teacher" at the beginning of the sections on "Animals," "Growing and Changing: Animal Life Cycles," and "Animal Adaptation."
- *Procedure*: The steps for the students to follow are given in easy-to-understand language.
- *Teacher Instructions:* Suggested teaching tips and background information follow the procedure. This information supplements that given in "To the Teacher."

Use of Metric Measures

Most linear measures used are given in metric units followed by standard units in parentheses. This is done to encourage use of the metric system. Other measures, such as capacity, are given in standard units.

Grade Level

The activities in this book are intended to be nongraded. Many activities in each section can be adapted for use with young children, yet most sections provide challenge for the more talented in the middle and upper elementary grades.

Final Note

Remember, the discovery/inquiry approach used in *Life Sciences* emphasizes verbal responses and discussion. With the exception of recording results, activities do not require extensive writing. Discovering the excitement of science and developing new techniques for critical thinking and problem solving should be the major goals of your science program.

Marvin N. Tolman
James O. Morton

ACKNOWLEDGMENTS

Mentioning the names of all individuals who contributed to the *Science Curriculum Activities Library* would require an additional volume. The authors are greatly indebted to the following:

- Teams of graduate students at the University of Utah for initial assistance and testing of design and methodology.
- Teachers and students of all levels, from early childhood through postdoctoral, who taught us while they thought we were teaching them.
- School districts throughout the United States who cooperated by supporting and evaluating ideas and methods used in this book.
- Friends who provided special facilities so we could have the freedom to work and create: Drs. Richard and Virginia Ratcliff and family, Mike and Sue Grinvalds and family, and Sally Gross.
- Special consultants who made significant contributions to the development, quality, and accuracy of the manuscript: Lornel T. Morton, Klamath Falls, Oregon (illustrating, typing, and editing); Sue P. Grinvalds, Klamath Falls, Oregon (illustrating); Gregory L. Tolman, Spanish Fork, Utah (illustrating); Bonnie R. Newman, Salt Lake City, Utah (consultant); Kathryn L. J. Ardt, Klamath Falls, Oregon (consultant); Vera A. Christensen, Logan, Utah (media); and Denise Swift (typing)
- Dr. James E. Baird, chairman of the Department of Elementary Education at Brigham Young University, for his encouragement and support, and for running interference to protect precious writing time.
- Finally, the last names of the authors, Tolman and Morton are correct. However, the first names could well be changed to Judy and Lornel, for without their love, support, encouragement, patience, and acceptance these books could not have been written.

CONTENTS

CONCEPT/SKILLS INDEX FOR BOOK I: LIFE SCIENCES

The following concept/skills index represents the scientific concepts or Big Ideas included in this book. The recommended age and grade levels at which the Big Ideas are introduced or developed in greater depth in each section are broken down this way:

preK: ages 3–5 (early childhood)
K–3: ages 5–8 (kindergarten-primary)
4–8: ages 9–14 (upper grades)

Each concept has been stated in simple but scientifically accurate terms followed by the numbers of recommended activities that introduce or reinforce the concept.

In some sections, background concepts necessary to an understanding of observed behavior cannot be "discovered" by hands-on experiences or activities at the elementary school level. When these Big Ideas are introduced, the activity column of the index will list "Teacher-student/demonstration/discussion." In these situations, strategies for presenting the necessary background information will be found in "To the Teacher" at the beginning of the section and "Teacher Information" after each activity.

As worded, the concepts/skills in this index are intended for teacher use, to give an overview of each section. They are *not* intended to be used as an evaluation of student knowledge.

Throughout the book, a hands-on, activity-centered approach is emphasized. The scientific process skills of observing, classifying, communicating, measuring, inferring, and experimenting are intended to be the major outcomes of this program. Content acquisition and development in science and other subject areas will be major concomitant learning. In attempting to recommend age or grade levels at which concepts should be introduced, we recognize that a great variation will exist because of individual differences in students which only you can determine.

Remember, a positive attitude toward science and "sciencing" on the part of both teachers and students is the major purpose of this book and will determine its effectiveness.

PLANTS AND SEEDS

Big Idea	*Suggested Grade Level*	*Activities*
Many plants grow from seeds.	preK–8	13
When exposed to moisture and warmth, most seeds begin to sprout. This is called germination.	preK–8	14

PLANTS AND SEEDS (continued)

Big Idea	*Suggested Grade Level*	*Activities*
Sprouts from most seeds grow downward due to geotropism.	preK–8	14
Many seeds have a baby plant and food within their outer covering. They do not need sunlight and soil until they begin to sprout green leaves.	preK–8	14, 15
As plants grow larger, their underground root systems, which take in food and water, must grow larger also.	preK–8	16
Water and food travel up the stem into the leaves of many plants.	preK–8	17
Plants and their containers can add beauty to our lives.	preK–8	18, 28
Some plants can be grown from the root parts of older plants of the same variety.	preK–8	19
Some plants can be grown from a "cutting" of an older plant of the same variety.	preK–8	20
As many plants grow older, they need additional amounts of soil, water, and light.	preK–8	21, 30
Many common plants have roots, stems, and leaves.	preK–8	22, 23
Leaves of green plants contain chlorophyll which, in the presence of sunlight, enables them to manufacture food. During this food-making process, called photosynthesis, plants also produce oxygen.	4–8	23 (Teacher-student/ demonstration/ discussion)
During respiration, a food-using process, plants use oxygen and give off carbon dioxide.	4–8	23 (Teacher-student/ demonstration/ discussion)

PLANTS AND SEEDS (continued)

Big Idea	*Suggested Grade Level*	*Activities*
Green plants produce more oxygen during photosynthesis than they use during respiration. They use more carbon dioxide during photosynthesis than they produce during respiration. Green plants are therefore an oxygen source. Respiration continues day and night. Photosynthesis takes place in the presence of light.	4–8	23 (Teacher-student/ demonstration/ discussion)
Flowers are important for the reproduction of many plants.	4–8	24
Complete flowers have four sets of parts: sepals, petals, stamens, and pistil.	4–8	24
Incomplete flowers have one or more of the parts missing.	4–8	24
Many plants will grow in the direction of a light source. This attraction to light is called phototropism.	K–8	25
Some plants use other plants or parts of once-living materials as a source of nutrients needed for manufacturing food.	4–8	26, 27, 28
Molds are unusual plants, with many of the following characteristics: a. Usually have a cottony texture b. Grow best in warm, dark, damp places c. Often damage food, leather, cloth, and paper d. Can cause diseases in people, plants, food, crops, and animals e. Help by causing wood, leaves, and other materials to rot, forming humus that makes soil rich	K–8	26, 27, 28, 29

PLANTS AND SEEDS (continued)

Big Idea	*Suggested Grade Level*	*Activities*
f. Are used by people to produce medicine (penicillin)		
g. Give off carbon dioxide that plants use to produce oxygen		
h. Reproduce by releasing spores that are carried through the air or by animals		
Seeds travel and are scattered or dispersed in many ways. Many travel through the air on fluffy cotton balls or wings. Some are dispersed by falling or are spread by animals. Sometimes a plant will even break off and "tumble" in the wind to scatter its seeds.	K–8	31
We eat many parts of plants.	K–8	33, 34
Plants provide the essential nutrients for life.	K–8	33

ANIMALS

Big Idea	*Suggested Grade Level*	*Activities*
Humans keep many different kinds of animals as pets.	preK–8	36
Different pets need special kinds of care, including:	preK–8	37
a. Type of living space		
b Type and amount of food		
c. Type of environment (water, dirt, rocks, grass, and so on)		
d. Amount of water and how consumed		
e. Living arrangement (alone, same species, same sex, mixed sexes, mixed species, and so on)		
f. Temperature requirements		
g. Amount and kind of exercise		
h. Amount of touching or holding		

ANIMALS (continued)

Big Idea	*Suggested Grade Level*	*Activities*
People keep pets for many reasons. Sometimes they do not or cannot care for the pets properly.	preK–8	37, 38, 39
Pet ownership includes a responsibility to give the pet proper care and to be conscious of the welfare of other people.	preK–8	37, 38, 39, 40
The hair on your head and the fur of some animals is similar in many ways.	preK–8	41
Many people of all ages like to own objects that feel and look like real or imaginary animals.	preK–8	42
You can create your own imaginary animals from paint, bones, and many other materials.	preK–8	43, 44
Every species of animal makes its own kind of track or footprint.	preK–8	45
You can learn many things about an animal by observing its tracks.	K–8	46, 47, 48
Many animals live near you.	4–8	48
You can learn to locate, observe, and study your animal "neighbors."	4–8	49, 50, 51
Scientists use special equipment and materials to help them study animals.	4–8	49, 50, 51
Scientists use many ways to report and share their findings.	4–8	52
Because they are so plentiful and different, wild birds are very interesting to study.	K–8	53
Wild birds will come near your home or school if you know how to attract them.	K–8	53
Birds have some characteristics that no other animals have.	K–8	54
You can make houses to attract different kinds of birds.	K–8	55
If you plan a special environment called an aquarium, fish and snails can live in your home or school.	K–8	56

ANIMALS (continued)

Big Idea	*Suggested Grade Level*	*Activities*
Small land animals can be housed in your home or school in an environment called a terrarium.	K–8	57
You can make different kinds of terrariums to meet the requirements of different types of animals.	K–8	57
Many animals, too tiny to study with the naked eye, live around us.	K–8	58
Earthworms are interesting, unusual, and helpful animals.	K–8	59
Snails are interesting to observe and many varieties are helpful to people.	K–8	60

GROWING AND CHANGING: ANIMAL LIFE CYCLES

Big Idea	*Suggested Grade Level*	*Activities*
Humans grow and change as they age.	preK–8	61
Humans go through at least four or five life stages.	preK–8	61
The young of humans and other mammals grow from fertilized eggs inside the female and are born in a developed state.	preK–8	62
When compared to most other animals, the young of mammals are quite helpless and require care for longer periods of time.	preK–8	62
Some animals lay eggs that, when fertilized, develop and hatch outside the mother's body.	preK–8	63, 65
Many varieties of animals hatch from eggs.	preK–8	64, 65
Fertilized eggs laid and hatched outside the female's body contain the food and materials necessary for the egg to develop into young animals.	preK–8	64, 65
Some animals resemble the adult of the species when they hatch; some do not.	preK–8	64

GROWING AND CHANGING: ANIMAL LIFE CYCLES (continued)

Big Idea	*Suggested Grade Level*	*Activities*
Some animals go through several changes before they become adults. This process is called metamorphosis.	preK–8	64
Amphibians, such as frogs, hatch from eggs and spend their first stage as a fishlike creature. Gradually they develop legs, arms, and lungs to become land-living animals.	K–8	66, 67
Many insects go through a complete metamorphosis from egg to larva to pupa to adult.	K–8	68
In the larva stage, some insects resemble worms, often called maggots.	K–8	69
Aging is a natural life process.	4–8	61, 70
Some animals complete their life cycles in a short time. Others have a longer life span.	4–8	61, 70

ANIMAL ADAPTATION

Big Idea	*Suggested Grade Level*	*Activities*
Animals, including humans, adapt in many ways.	preK–8	71
Environment is a major factor in animal adaptation.	preK–8	71
Some animal adaptations are highly specialized.	preK–8	72
People often imitate special animal adaptations but cannot always improve upon them.	K–8	73, 74, 84
Color, or its absence, is an important adaptation of many animals.	K–8	75
We use color in many ways in our lives.	K–8	75
There is a relationship between their mouthparts and the type and variety of food animals eat.	K–8	76, 77

ANIMAL ADAPTATION (continued)

Big Idea	*Suggested Grade Level*	*Activities*
Human mouthparts are adapted for eating a wide variety of foods.	4–8	78
Because of recent changes in food habits, human mouthparts require additional care.	4–8	79
Animals move in many ways.	K–8	80
The environment in which animals live affects their type of movement.	K–8	80
Many animals communicate in different ways.	K–8	81
Complex symbolic language is peculiar to humans.	4–8	81
Seasonal changes influence the adaptations of many animals.	4–8	82, 83

BODY STRUCTURE

Big Idea	*Suggested Grade Level*	*Activities*
People grow at different rates and to different sizes.	K–8	85
The organs and systems of the body work together.	4–8	86
Lung capacity usually increases with proper exercise.	K–8	87
Our fingernails grow rapidly all the time.	K–8	88
Fingerprints are unique for each individual.	K–8	89
Pigment determines hair color and is contained in the middle of three layers of human hair. As people grow older, the pigment sometimes disappears and the hair turns white.	4–8	90
The skin helps to protect the body and helps regulate body temperature.	4–8	91, 92
Bones provide structure for the body, allow it to bend, and protect the vital organs.	4–8	93

BODY STRUCTURE (continued)

Big Idea	*Suggested Grade Level*	*Activities*
Human bones are much larger than those of some animals but much smaller then others'.	K–8	94, 100
The human body has a total of at least 206 bones.	K–8	95, 100
By using X-rays, a doctor can see the bones and determine their state of health.	4–8	96, 97
Bones are connected at joints by tendons.	4–8	98, 99
The human body has over 650 muscles, including several different types.	4–8	101
Voluntary muscles, such as the arm muscles, are controlled at will.	4–8	102, 103

THE FIVE SENSES

Big Idea	*Suggested Grade Level*	*Activities*
Pictures of things can stimulate other senses.	K–8	104
Odors disperse through the air in all directions from their source.	K–8	105, 106
The sense of smell becomes less sensitive to a given odor if exposed to it for an extended period of time	4–8	107
Foods do not taste exactly the same to everyone.	K–8	108
Tastes are classified into four categories: sweet, sour, salty, and bitter.	4–8	108, 112
Taste is affected by the sense of smell	K–8	109, 111
Certain parts of the tongue are more sensitive to certain tastes.	4–8	110
The sense of hearing helps to keep us aware of what is happening around us.	K–8	113, 114, 115, 116, 117, 118
Peripheral vision is the ability to see at the side while focusing straight ahead.	4–8	119

THE FIVE SENSES (continued)

Big Idea	*Suggested Grade Level*	*Activities*
With practice, visual memory skills can be improved.	4–8	120
Our eyes can help us to coordinate and add information to our other senses.	4–8	121, 122
The iris of the eye opens and closes like the diaphragm of a camera, according to light intensity. This change is visible as it alters the size of the pupil.	4–8	123, 124
The two eyes, being a short distance apart, see objects from a slightly different angle. This aids perception of depth and distance.	4–8	125
Most people have one eye that is dominant and determines line of vision.	4–8	126
The five senses interact and assist each other in greater sensitivity and accuracy.	4–8	127, 130
Touch can help us gain additional sensory information.	K–8	128, 131, 132
The sense of touch is more sensitive on some areas of the skin than others due to varying concentrations of nerve endings and receptors.	4–8	129, 133

HEALTH AND NUTRITION

Big Idea	*Suggested Grade Level*	*Activities*
Good sanitation habits promote good health.	preK–8	134
A balanced diet consists of appropriate servings of each of the four basic food groups.	K–8	135, 137
Popular foods are frequently not nutritious.	K–8	136
Protein is a body builder. Tests show its presence in food.	4–8	138

HEALTH AND NUTRITION (continued)

Big Idea	*Suggested Grade Level*	*Activities*
Limiting the amounts of sugar, starch, and fat consumed promotes good health. There are ways to find out how much fat, sugar, and starch you eat.	4–8	139, 140, 141
The calorie is the unit of measure of the amount of energy of food will produce when digested in the body.	4–8	142, 143
Proper balance between the intake and use of calories promotes good health.	4–8	142, 143
The different types of teeth are designed for specific tasks.	K–8	144
Maintaining healthy teeth requires proper care.	K–8	145, 146

Section 1

STARTER IDEAS

TO THE TEACHER

The following section is a challenge to you. The ideas presented here have been collected over a period of many years from the bright, creative minds of thousands of scientists, teachers, students, and children of all ages. Within your mind are creative ideas that we hope these Starter Ideas will stimulate. The challenge is an invitation to explore, inquire, invent, and create.

We hope that somewhere in these simple, easy, inexpensive activities you will find your special invitation for creative inquiry. With few exceptions, the sections of this book do not build sequentially, so you can begin almost anywhere and explore forward or backward.

Starter Ideas are single-concept ideas that range far and wide in the unlimited areas we often try to cover in science. (One dictionary definition of the term *cover* is *to conceal.* We hope you will not try to conceal these ideas but instead will uncover, explore, and develop them.) Starter Ideas are intended to start your exploration in the exciting world of the sciences.

Try the ideas at random. When you find something you and your students like, there will usually be a reference to the section of the book where you can explore it in greater depth. If you or your students become bored or don't like an area, think of ways to add interest or move to another science topic. You needn't do everything just because you enjoy part of it. As educators, we know young people do best the things they enjoy most. Your best teaching is done for the same reason. If you find something you don't understand, look for new creative ways to learn and develop the concept. New interests will blossom as you explore new horizons and student interest is likely to increase along with yours.

If there is one "must" everyone in our society needs to understand, it is that we have unlimited wants and limited fragile natural resources. It is important that we use what we have wisely. We must also understand that science, especially the new technology, must remain our servant, not become our master.

Whatever you choose, an activity is in this first section, "Starter Ideas," to help you get started. The purpose is to explore and inquire creatively and through these explorations develop every facet of your students' ability to learn.

In this section there is really nothing you have to do but explore, enjoy, and create. Somewhere in the next few pages you will find the place to begin your exciting journey in science.

There is an ancient Chinese saying: "A journey of a thousand miles begins with a single step." It is our hope that the Starter Ideas will provide that first step.

ACTIVITY 1: Can We Communicate with Plants?

(This is a "just-for-fun" activity)

MATERIALS NEEDED

- Six to eight house plants of the same variety, age, and vigor
- Sensitive scale to weigh grams and ounces (food scale)
- 9″ × 12″ newsprint
- Markers
- Player and records or tapes of different kinds of music
- Name tags

BACKGROUND INFORMATION

Scientists have long wondered if there might be a way to communicate with plants. Although there has never been any scientific evidence that plants can talk or hear, you may be communicating with them in special ways through the care you give them. Most plants need sunlight, water, warmth, and the correct kind of soil. Do they need or want anything else? Just for fun, try some of these things. As you begin, you will probably think of many more ideas to try.

PROCEDURE

1. Make pictures and describe each plant. Measure them. Weigh them and record.
2. Name all but one of your plants and attach the names to their containers. When you come near them to water or move them always use their names and ask them how they are feeling. Care for the plant without a name in exactly the same way but don't talk to it.
3. Put two of your plants with names closer together. Isolate the plant without a name.
4. Play soft soothing music to two of your plants daily for one hour. Play loud harsh music to two others.
5. Try other ideas on different plants, but remember to do just one thing different to any plant. Keep a record of what you did to each plant.
6. After two or three weeks, bring all your plants together. Observe them. Make pictures of them. Weigh them and measure them again. What can you say about this experience?
7. Remember, this is not a scientific experiment. It's just for fun.

TEACHER INFORMATION

Although there is no scientific evidence that plants communicate with humans or each other in a linguistic sense, there is some evidence to suggest that certain varieties of plants (trees) may send out a message to others of the same variety when they are under attack by insects or fungus.

People who love plants may ascribe personalities to them, talk to them, and give them better care, and the plants will thrive. This could lead to an interesting discussion of how plants should be treated in our society.

This activity may also be found in Section 2, "Plants and Seeds."

ACTIVITY 2: What Can Hair Tell You?

MATERIALS NEEDED

- Large, gentle dog
- Newspapers
- Markers
- Clean, wire pet brush
- Small plastic bags
- Small labels

PROCEDURE

1. Have your teacher help you find a large, very gentle, well-trained dog with a shaggy coat. A dog who lives outside most of the year is best. Do *not* use a stray.
2. Spread newspapers on the floor. Use a clean, wire pet brush to brush your dog ten times (strokes). Remember the exact area you brushed. You might want to make a drawing of the dog to help you remember.
3. Collect all the dog hair from the newspaper and wire bush. Store the hair in a small plastic bag. Put a tag on the bag telling the day, date, and time of the brushing.
4. Brush your dog in exactly the same place, in the same way, four times during the school year. September, December, March, and late May might be best.
5. Examine and compare the hair you collected after each brushing.
6. What can you say about this?

TEACHER INFORMATION

This activity should be approved by your principal. Before you brush the dog, it might be well to have it visit several times to get acquainted with the class. This activity is most effective in climates with pronounced seasonal changes. However, most dogs will undergo seasonal changes in the thickness of their coats, even in mild climates.

Animals with hair shed constantly (humans included). Animals who normally live outdoors often develop a thick undercoating in winter. The condition of the hair of a mammal is often an indicator of the animal's health (and of humans).

Students might notice that they lose hair regularly. Unless the loss of hair is extreme or in patches it is part of the normal hair growth pattern. At any one time about 20 percent of human hair is in the "resting" stage; that is, the hair is gone but the follicle will remain, and in a few months will grow a new hair. Baldness, which occurs mostly in males, is hereditary and so far is irreversible, though remedies are sought.

The fur of many mammals serves the dual purpose of warmth and protection. Hair in humans is mostly for protection. Hair on our head insulates somewhat, but mostly protects the skull. Eyebrows shade our eyes from the sun. Eyelashes and hair in the nose and ears keep dust and insects out. Some scientists believe the tiny sensitive hairs around our months serve the same purpose as feelers in insects.

All hair is not the same shape. If you examine different types of hair under a low-power microscope or magnifying glass you will find that straight hair is round. Curly hair is flat.

This activity is also included in Section 3, "Animals."

ACTIVITY 3: Can You Save the Fish?

MATERIALS NEEDED

- Pencil with eraser
- Prepared index card
- Pin
- Glue

PROCEDURE

1. Hold the pencil between the open palms of your hands.
2. Move your palms back and forth rapidly so the pencil spins around.
3. Observe the fish on the card. What happened?
4. What can you say about this?

TEACHER INFORMATION

Fold and cut an index card so you have a two-inch square on two sides. Draw a fish on one side of the card and a bowl on the other. Glue the card together with the pin in the middle. Stick the pin in the eraser of a pencil. When the pencil is rotated in the palms of the hands, the fish will appear to be in the bowl. This demonstrates the idea of persistence of vision. When we see an image it persists for about 1/16 second. If another image appears within that time, we will see both. Thus the fish appears to be in the bowl. Other related objects such as a lion and a cage or a basketball and a basket can be substituted.

Because of this phenomenon, if approximately 24 pictures move in front of the eye each second, they can change slightly and blend together into a "moving picture."

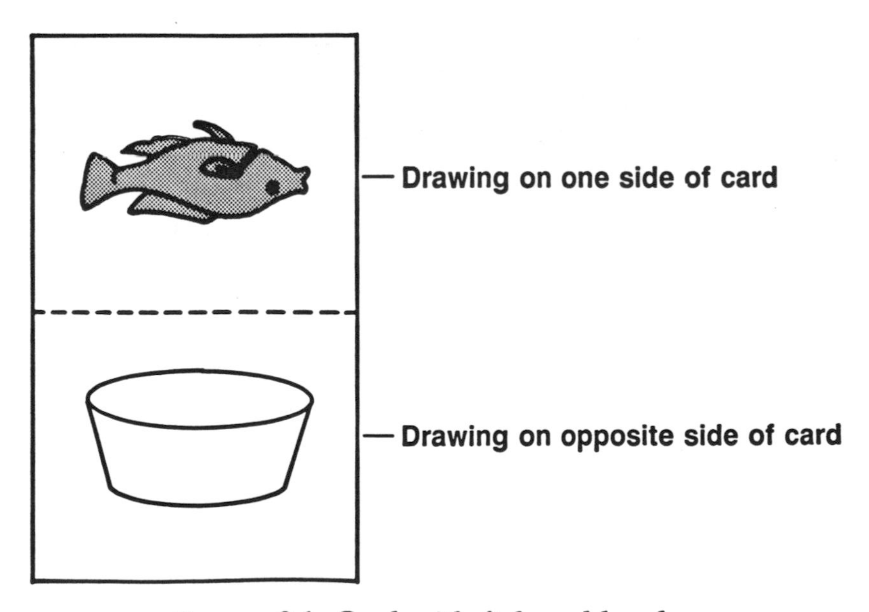

FIGURE 3-1. Card with fish and bowl.

ACTIVITY 4: What Is a Blind Spot?

MATERIALS NEEDED

- Prepared 5″ × 7″ index card
- Broad black felt-tipped pen

PROCEDURE

1. Close and cover your left eye.
2. Hold the index card at arm's length.
3. Stare at the X.
4. Slowly bring the card closer to your right eye.
5. What happened to the dot? Repeat this activity several times.
6. Discuss this with your teacher and the class.

TEACHER INFORMATION

As the student slowly brings the card near while staring at the X, at some point the dot will momentarily disappear. This is because each eye has a blind spot where the optic nerve exits the eye. At this point there are no rods or cones to project an image. The blind spot occurs only at a specific distance. We usually see beyond or within this distance, and we use two eyes, so we usually don't notice the blind spot. Have students experiment, varying the size of the X and the dot and varying the distance between them.

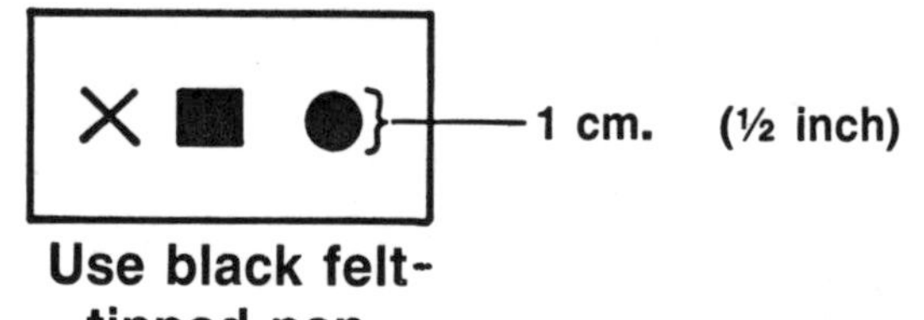

FIGURE 4-1. Index card with X and dot.

ACTIVITY 5: How Can Our Eyes Appear to Make Colors Change?

MATERIALS NEEDED

- Prepared 5″ × 7″ index card
- Broad blue and red felt-tipped pens

PROCEDURE

1. Stare at the blue and red squares for 30 seconds.
2. Now stare at the flat white surface.
3. What happened? What can you say about this?

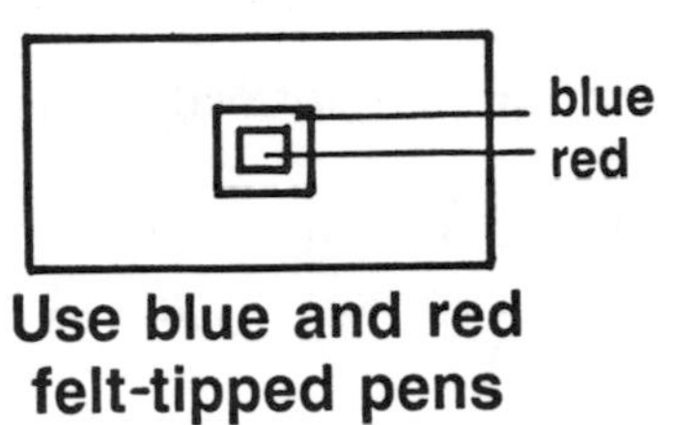

FIGURE 5-1. Index card with red and blue squares.

TEACHER INFORMATION

A simple explanation for elementary children might be: When we stare at bright colors for a length of time, the cones in our eyes that see the bright colors get tired. When we look at a white surface, the tired cones rest and the other cones near the same place in the eye take over. We will still see the image we have been staring at but it will be in different colors. (This may be a place to explore the idea of complementary colors in art.)

A related idea is the afterimage effect. Darken the room and stare at a vivid object for several seconds (try a large black X on a white sheet of paper). After the object is removed you will still be able to see it faintly; however, the colors will be reversed (white X on a black square) just as in a negative of a black-and-white photograph.

ACTIVITY 6: How Does a Sudden Change in Temperature Affect Us?

MATERIALS NEEDED

- Large can or bowl filled with hot water
- Large can or bowl filled with warm water
- Large can or bowl filled with cold water

PROCEDURE

1. Observe the three cans.
2. One has hot water in it.
3. One has warm water in it.
4. One has cold water in it.
5. *Carefully* put one hand in the hot water and one in the cold. Leave them there for 30 seconds.
6. Now put both hands in the warm water.
7. What happened? What can you say about this?

TEACHER INFORMATION

When placed in the warm water, the hand that was in the hot water will feel cool while the hand from the cold water will feel warm. A simple explanation for elementary children is that our body has a control system to help us adjust to hot and cold. After one hand has adjusted to hot water and the other has adjusted to cold water, the warm water will feel quite cool to the one and quite warm to the other. This happens because of contrast with that to which they have become accustomed. This is the reason a room feels hot when we first come in from the cold outdoors or vice versa.

ACTIVITY 7: How Can You See Through a Solid Object?

MATERIALS NEEDED

- Cardboard tube, such as a toilet tissue tube or paper towel tube
- Book

PROCEDURE

1. Look through the cardboard tube at an object across the room. Keep your other eye open, too.
2. While staring at the object with both eyes, bring your open hand (or a book) against the side of the tube near the far end so the vision in the eye not looking through the tube is blocked (remember to keep both eyes open).
3. What happened? Discuss this with your teacher.

FIGURE 7-1. Student looking through cardboard tube.

TEACHER INFORMATION

Your hand or the book, when brought against the side of the tube, will appear to have a hole in it. You will see the object farther away through the hole. Those of us who are fortunate enough to have two eyes have two receptors sending images to the brain simultaneously. The brain combines the two images and the distant object seems to be seen through a hole in your hand.

ACTIVITY 8: What Is a Stethoscope?

(Teacher-supervised partners or small groups)

MATERIALS NEEDED

- Stethoscopes (at least for each two or three participants)
- Soft facial tissue or toilet paper

PROCEDURE

1. Stethoscopes are delicate instruments especially designed to help us *listen* to functions inside our bodies. Examine your stethoscope. Notice that it resembles the letter Y.
2. Can you find the two small rounded ends, held together with a spring? They fit in your ears.
3. Use a soft tissue to clean the rounded ends and carefully put one in each ear. The spring should keep the ends in place. Be sure to clean the ends before each use.
4. Notice that the two hollow tubes coming from the ends in your ears join together to make a single tube.
5. The single tube is attached to a large, flat disc. Carefully examine the flat disc. It has a sensitive vibrator or *diaphragm* that magnifies sound.
6. Put the disc against the part of your body where you think your heart is.
7. Use the stethoscope to listen to other parts inside your body: stomach, throat (try swallowing), lungs (breathe deeply and listen both front and back). Try other places.
8. A stethoscope helps our hearing in the same way magnifying glasses help our vision. Can you explain how?

TEACHER INFORMATION

Demonstrate first for younger children. Stethoscopes are available from medical personnel—doctors, nurses, dentists, and medical technicians. Some students may have them in their homes. The diaphragms are delicate and should be handled with care.

Some students are familiar with stethoscopes, but many are not. Students can hear many of the normal functions in their own bodies, and may be surprised at the normal sounds (such as those in the intestinal tract) that go on regularly inside them. Listening to throats, hearts, and breathing may require partners or teams. Modesty should be observed.

Stethoscopes can also be used to hear the small sounds around us (for example, small insect noises). If possible, have one available at all times for students to use.

Modern technology may make the stethoscope obsolete. Electronic devices are gradually replacing it in many fields of medicine.

For additional activities, see Section 6, "Body Structure."

ACTIVITY 9: What Plant Is This?

(Advance teacher preparation needed)

MATERIALS NEEDED

- Variety of seeds from common garden plants, such as radish, corn, carrot, onion, beans, and other seeds found in your region (see "Teacher Information")
- Small container such as paper, foam, or plastic cup (one per student) filled with moist potting soil containing two or three seeds from *one* variety of plant listed above
- Library books with pictures of the plants that will grow from the seeds
- Masking tape
- Pencils
- Water

PROCEDURE

1. Your container has potting soil and seeds of a plant in it. Write your name on a piece of masking tape and put it on your container.
2. Put it in a warm, shady place in the room.
3. Give it a *small* amount of water daily.
4. In a few days a tiny, green plant will come up.
5. Move your plant into a sunny spot in the room. Continue to water sparingly.
6. As your plant grows larger, compare it with the plants of everyone else in the room.
7. Try to find everyone who has a plant that looks like yours. Form a study group.
8. With the other members of your group, use library books to find a picture of your plant. (No fair pulling it up to look at the roots!)
9. When you think you know the name of your plant, consult your teacher to see if it is correct.

TEACHER INFORMATION

This activity will cover a time span of several weeks. Other science activities can be done along with it. The introductory activities in Section 2, "Plants and Seeds," could parallel this activity.

Students may enjoy having a few surprise seeds, such as popcorn (unpopped) or raw peanuts. If you live in an area where a particular plant (corn, wheat, peanuts, etc.) is a common agricultural crop known to the students, omit it.

If space or time is limited, you could have several small plants with incorrect labels on them and ask the students to help you learn about the plants and change the names around until they match the plants correctly.

ACTIVITY 10: What Are We Made Of?

(Teacher demonstration in a sink, large bucket, or dishpan)

MATERIALS NEEDED

- Small 20-40 liter (5-10 gallon) *opaque* garbage bag
- Water
- Food products in small quantities such as hardboiled egg; milk; green, leafy vegetables; yellow vegetables (squash); piece of whole wheat bread; soy or any dry beans; potato with skin left on; small pieces of fruit—orange, apple, etc.; nuts
- Slice of pizza (avoid fresh meat and fish)
- Flashlight battery
- Tablespoon of salt
- Pinch of dirt

PROCEDURE

1. Add water to the bag until it is about three-fourths full.
2. As you add small quantities of the other materials in the list, discuss the importance (to human health and nutrition) of each item.
3. Continue to add items (especially leafy vegetables and fruit) until the bag is full.
4. Seal the top of the bag. Carefully shake the contents.
5. Although the amounts are not exactly in proportion, this bag is the way a chemist might see us. In order to grow, we must continue to add nutritional foods and water to our bag. What organ of the body does our bag represent?

TEACHER INFORMATION

You can keep costs down by using small quantities and perhaps asking your produce department at the local grocery store for scraps.

This simple idea may help introduce the following concepts:

a. Our body is mostly water (over nine-tenths).
b. Various foods contain different amounts of elements our body must have.
c. A balanced diet means giving our bodies the right kinds of quantities of many foods.
d. One important purpose of our skin is to keep our body contents in and many other things out. In our model, the garbage bag represents the skin, which is the largest organ in the body.

The salt and dirt represent bacteria and compounds that are natural and essential to body functions. In Section 8, "Health and Nutrition," you may want to discuss health problems that require limitations in salt and other substances and the difference between harmful and helpful bacteria in the human body.

The flashlight battery is optional and probably should be used only with older students, although younger ones may relate to the concept of energy in the body. It represents the chemical reactions that produce electrical energy, from which our brain and nervous system operate.

Life processes such as digestion, respiration, and elimination are not included in this simple model. It might, however, be the model that could trigger many creative and challenging questions such as: "Now that we have all the necessary chemical contents in this bag, why isn't it alive?" or "What does your body do that this bag cannot?"

ACTIVITY 11: How Can You See Your Pulse?

MATERIALS NEEDED

- Large thumbtacks
- Used wooden kitchen match

PROCEDURE

1. Carefully insert the point of the thumbtack into the end of a wooden match. The match should extend vertically from the thumbtack.
2. Rest your hand, palm up, on a flat surface.
3. Using the thumbtack as a base, place the match in an upright position on your wrist.
4. Move the match to different positions and observe it.
5. What happened? Why?
6. Run in place for one minute.
7. Repeat the activity. Was there a difference?

TEACHER INFORMATION

Most students have had their pulse taken at some time. Often they don't understand why or what the doctor or nurse is doing.

A wooden match placed on the wrist, especially in an area in approximate line with the index finger, should move back and forth noticeably.

The pulse is difficult to locate in some people. In some cases medical personnel may use the throat. These differences are normal.

Pulse rate varies from individual to individual and within an individual at different times. Related activities on circulation may be found in Section 6, "Body Structure."

ACTIVITY 12: What Is a Blood-Pressure Cuff?

(Teacher demonstration and supervised partners or small groups)

MATERIALS NEEDED

- Blood-pressure cuff with gauge

PROCEDURE

1. Examine the scientific instrument. It has a very long name: *sphygmomanometer.* Most people call it by its common name: blood-pressure cuff. Most of you have seen an instrument similar to this. Medical personnel use it to help them *monitor* (watch) your heart and circulatory system. Identify the cloth cuff, the rubber bulb and tube, and the gauge.
2. Locate the screw nob below the gauge. It regulates the pressure (amount of air) in the cuff. The rubber bulb will pump air into the cloth cuff. By turning the knob, you can control the amount of air in the cuff.
3. Lay the cuff flat on a table. Use the bulb and knob to practice pumping and releasing air from it. *Don't pump it too full or you may damage it.*
4. Study the gauge. It has a dial with numbers on it and a needle that tells the amount of pressure you have pumped into the cuff. Practice using and reading the gauge.
5. Have your partner wrap the long, wide strip of cloth around your upper arm (above the elbow) and use the snap clips or velcro to fasten it. Be sure the gauge and bulb are on the outside so you can see and touch them.
6. With a partner, practice pumping and releasing air from the cuff on your arms. Be sure to watch the gauge. As the air is released, notice that you can feel a pumping sensation in your arm. Be careful not to pump the air in the cuff so tightly that it hurts. Begin to release the pressure at once. *Never leave a tight cuff in place on your arm.*

TEACHER INFORMATION

This activity should begin with a teacher demonstration and with careful supervision throughout. Circulation to the arm should not be cut off for more than a few seconds. If pressure in the cuff is too great, delicate blood vessels could be damaged.

The main objective of this activity is to show how scientists use instruments to help provide information they need. In some ways it is analogous to the gauges and dials in your automobile.

Some of the anxieties many of us have about going to the doctor can be reduced through an understanding of the why and how of many of the instruments used. Many pediatricians today understand this concept and attempt to allay children's fear. School activities such as this should help support the efforts of parents and medical personnel.

When students understand how a stethoscope and blood-pressure cuff work, invite a doctor, nurse, or medical technician to show how to take blood pressure and explain what it means.

This Starter Idea is related to activities found in Section 6, "Body Structure."

Section 2

PLANTS AND SEEDS

TO THE TEACHER

A study of plants affords many opportunities for creative imagination. Most activities are nongraded, but it is recommended that they be taught sequentially from seed to plant. Throughout the study, appreciation of beauty and the wonder of growth should be emphasized. Language arts, music, art, and other subjects should be related as often as possible.

Teachers in urban areas may find some difficulty in conducting a few of the field trip activities; however, window boxes and artificial growing areas may be substituted.

Resource people can be valuable in this area. Most classrooms have mothers or fathers who grow plants as a hobby. Agronomists (soil specialists), horticulturists, florists, botanists, and nutritionists can also be helpful in their areas of specialization. Designing and planting a school garden could be a very worthwhile outgrowth of this study.

Throughout the study, care should be taken to emphasize the danger of eating any unknown substance. Even some parts of plants we eat can be poisonous, such as rhubarb leaves and some varieties of potato plant leaves. Warn children never to eat berries or flowers. On field trips, take into account regional variations to avoid such things as poison ivy and stinging nettle.

Most growing activities suggest the use of potting soil; however, as plants develop they will usually do better in rich, loamy soil. Specialized plants such as cactuses and pine trees will need special alkali or acid soils. Be sure to check reference sources before growing specialized plants.

Older children may be interested in solving a problem as a plant scientist might. Several plants in the same kinds of containers could be presented with the following statement and question:

> These are all the same species of plant, planted on the same day from similar seeds. How can you account for the differences?

The plants used will have been grown in different media—gravel, sand, potting soil, or loam. One will have been put in a dark place, one in sunlight. One will have been overwatered, one underwatered. As children observe differences in the conditions of the plants, soil, and moisture, they will form hypotheses they can test through replication: identifying a problem (some plants don't grow as well as others); stating a hypothesis (this plant is too dry); testing the hypothesis (if I transplant this plant to soil instead of gravel it will grow better); conclusion (what I tried worked or didn't work). This activity is similar to the work of agricultural specialists.

FINAL THOUGHTS

1. Memorizing scientific names and plant parts are not important at the elementary level.
2. Plant study affords many excellent opportunities for integration of other subject areas such as language arts, math, art, social studies, music, and health.
3. Our major purpose is to develop understanding, respect, and appreciation for the contributions plants make in our lives.

ACTIVITY 13: How Do Many Plants Begin to Grow?

MATERIALS NEEDED

- Large dry lima bean seeds
- Container of water
- Picture of lima bean seed split open

PROCEDURE

1. Observe the dry lima bean seeds.
2. Soak the seeds in water for 24 hours. How did they change?
3. Carefully split the seeds in half.
4. Try to find the baby plant and the other parts shown in the picture.

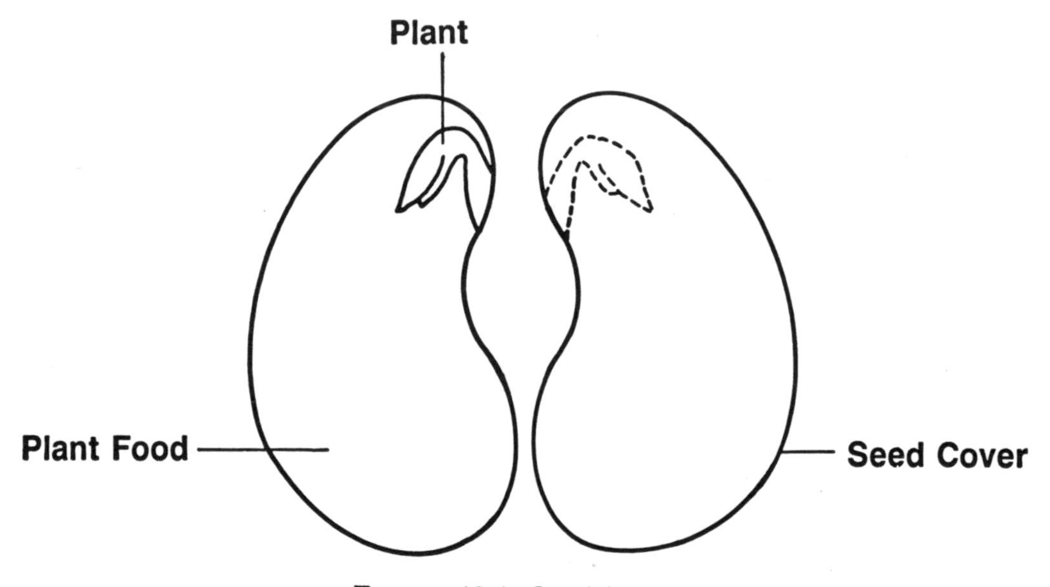

FIGURE 13-1. Seed halves.

TEACHER INFORMATION

Seeds that produce flowering plants come in many shapes and sizes. They all have three things in common: a protective cover, a food supply, and a baby plant called the embryo. Peas or raw peanuts may be substituted for beans.

This is an excellent individual or small group activity.

ACTIVITY 14: How Can We Watch Seeds Grow?

MATERIALS NEEDED

- Two 8-ounce clear plastic tumblers
- Two 10 cm. × 15 cm. (or 2″ × 4″) sponges
- Dry lima bean seeds
- Water

PROCEDURE

1. Soak the bean seeds for 24 hours.
2. Put a damp sponge around the inside surface of each tumbler.
3. Put eight seeds between the sponge and the side of the tumbler. The seeds should be placed 2 or 3 cm. (about 1 in.) apart.
4. Check the sponge every day. Be sure it is moist.
5. Put one tumbler in a dark location. Place the other in the light of the classroom.
6. Observe your seeds daily.
7. What happened? What can you say about this?

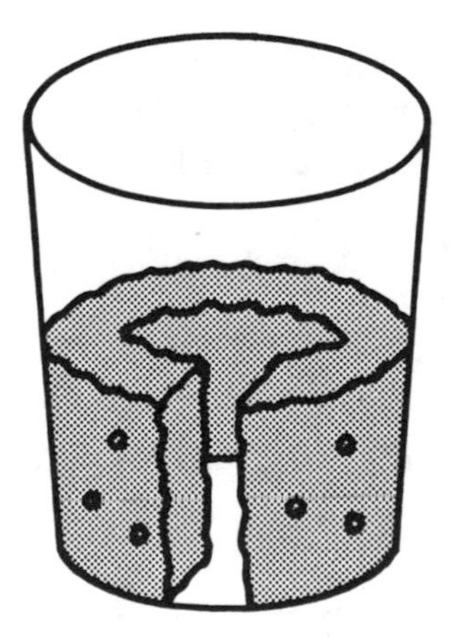

FIGURE 14-1. Seeds in a glass with sponge.

TEACHER INFORMATION

After several days the bean seeds will begin to sprout. The seeds kept in the dark place should sprout at about the same time as the ones in the light. Seeds do not need sunlight until they begin to grow leaves. Remind the children that many plants begin their lives in the dark, below the surface of the soil.

As the seeds continue to sprout, turn one tumbler on its side. The sprouts will change their direction of growth and grow downward. In whatever position the glass is placed, the sprouts will grow downward. This phenomenon is called *geotropism*.

ACTIVITY 15: What Do Plants Need in Order to Grow?

MATERIALS NEEDED

- One 4-ounce paper cup for each child
- Potting soil
- Lima bean seeds (same variety as those used in Activity 14)
- Water

PROCEDURE

1. Add potting soil until the paper cup is half full.
2. Plant a bean seed in the potting soil approximately one cm. (¼ in. to ½ in.) deep. Be sure to cover the seed.
3. Keep the potting soil moist and warm. Do not overwater.
4. Observe for several days.
5. What happened? What can you say about this?

TEACHER INFORMATION

Seeds kept moist at room temperature should begin to grow within three to five days. In about a week green sprouts will appear above the potting soil. Until green plants appear it is not necessary to keep the cups in sunlight (see Activity 14).

In place of bean seeds the children may want to try corn, peas, raw peanuts, wild bird seed, radish, geranium, or other seeds.

A suggested enrichment activity is to plant different seeds and measure the time of germination, sprouting, and rate of growth.

Seeds can be dormant for different periods of time and then begin to germinate and grow. Some can be kept only a few weeks, while others may germinate after 50 years. Under favorable conditions, scientists have been able to cause 10,000-year-old lotus seeds to germinate

ACTIVITY 16: How Do Roots Develop and Grow?

MATERIALS NEEDED

- Germinating seeds in a tumbler (prepared in the same way as in Activity 14)
- Sprouts in paper cups from Activity 15
- Magnifying glass, if available

PROCEDURE

1. Examine the roots of the seeds in the tumbler.
2. Pull up one sprouting plant from a paper cup and examine the roots (the seeds in the tumbler and the ones in soil in the paper cups should be of the same variety).
3. Compare the roots of the seeds in the tumbler with those of the sprouting plant in the paper cup. (You might need to use a magnifying glass.)

Time	Picture	Time	Picture

FIGURE 16-1. Chart.

4. Pull up a different plant every other day for a week and compare its roots to those of the ones in the tumbler. Record your observations on the chart.
5. What happened? What can you say about this?

TEACHER INFORMATION

Since the plants in the paper cups are more advanced, their root systems will be more highly developed. Small hair-like roots will be growing through the soil. During the week's observation, the roots from the seeds in the tumbler will begin to grow similar hairs. This activity could be continued for several weeks or lead to a study of other plants and their root systems.

ACTIVITY 17: How Does Water Travel in a Plant?

MATERIALS NEEDED

- Celery stalk
- Pint jar
- Water
- Red or blue food coloring.

PROCEDURE

1. Fill a pint jar with water.
2. Place a stalk of celery, leaves up, in the jar of water.
3. Cut one centimeter off the bottom of the celery stalk, under water.
4. Add four to six drops of food coloring to the water.
5. Observe for several days.
6. What happened?
7. What can you say about this?

FIGURE 17-1. Celery.

TEACHER INFORMATION

This activity will work best if the celery is cut under the water and *not exposed to the air.* In a few days the food coloring (red or blue works best), will travel up the celery stalk and color the celery leaves.

Note: This can also be done with a white carnation. One color can be used, or the stem can be split carefully and placed in two colors.

FIGURE 17-2. Carnation.

ACTIVITY 18: How Do Containers Make Plants More Interesting?

MATERIALS NEEDED

- Potting soil
- Various seeds
- Plastic egg cartons
- Eggshells
- Seashells
- Grapefruit or orange rinds
- Grapefruit or orange seeds
- Other assorted planters
- Water

PROCEDURE

1. Choose an attractive container.
2. Fill it with potting soil.
3. Plant a seed just below the surface of the soil.
4. Keep the soil moist and watch your plant grow.
5. If you choose a grapefruit or orange rind, you might want to try a matching grapefruit or orange seed.

FIGURE 18-1. Creative planters.

TEACHER INFORMATION

Seeds will sprout in almost any container that will hold potting soil. Objects that rust or are painted on the inside should be avoided. Plants in small containers, such as eggshells or egg cartons, will need to be transplanted. When this becomes necessary, simply crack the eggshell or cut through the carton and transplant into a larger container without disturbing the root system. This can be an opportunity to emphasize creativity and develop an appreciation for the beauty of plants. Any container that is to be used as a planter for an extended period of time must provide drainage.

ACTIVITY 19: How Can Plants Grow Without Seeds?

MATERIALS NEEDED

- Small aluminum pie tins or bowls
- Toothpicks
- 8-ounce plastic tumblers
- Freshly cut carrots, potatoes, sweet potatoes, onions, and beets
- Water

PROCEDURE

1. Choose a vegetable and a container.
2. Put your vegetable in the container exactly as shown in the picture, using toothpicks for support where necessary.
3. Add water to the container every day for at least two weeks.
4. What happened? What can you say about this?

FIGURE 19-1. Variety of plants.

TEACHER INFORMATION

This is an excellent individual activity. Be certain to have plenty of materials available so children can try more than one if they wish. Potatoes and sweet potatoes should have several eyes and be freshly cut. The green part of the carrot should not be submerged in water. Avocado seeds can be grown in the same manner as potatoes and will produce beautiful plants. It is important to remember, however, that the avocado is a seed, not another part of the plant.

Onions are bulbs (similar to tulips) and can be easily transplanted. The carrot is a root and the potato is called a tuber.

ACTIVITY 20: What Is Another Way to Grow Plants?

MATERIALS NEEDED

- Water
- Rubber band
- Pencil
- Knife
- 8-ounce glass tumbler
- Aluminum foil
- Philodendron plants

PROCEDURE

1. Fill the tumbler with water.
2. Cover the top of the glass with aluminum foil and hold it in place with an elastic band.
3. With your pencil, punch a hole in the center of the foil.
4. Cut a young, leafy stem, at least 20 cm. (8 in.) from the plant. (*Caution:* Don't break the stem off. Cut it cleanly.)
5. Trim the leaves from the bottom 10 cm. (4 in.) of the stem.
6. Insert the stem through the hole in the foil. Be sure the stem is in water. You might want to punch a second hole in the foil so you can add water.
7. Observe for at least one week.
8. What happened? What can you say about this?

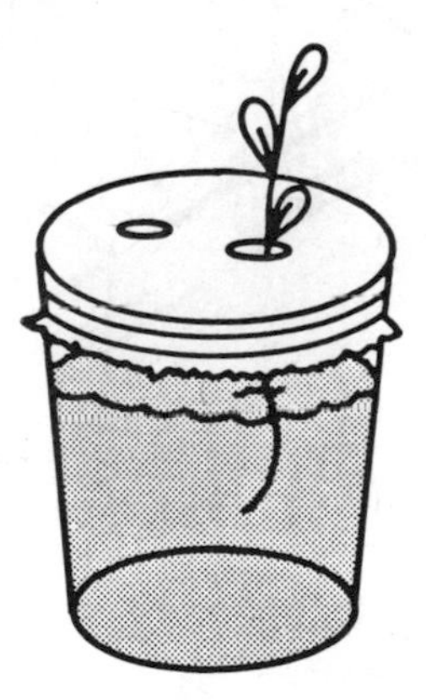

FIGURE 20-1. Plant cutting through foil and into water.

TEACHER INFORMATION

Keep the plant in sunlight. Many plants are started this way. The part cut away is often called a *start*, a clipping, or a cutting. The following activity will suggest ways to transplant the starts. Also, this can lead to a discussion of plant grafting. See your encyclopedia for information.

ACTIVITY 21: How Can You Keep Your Classroom Plants Growing?

MATERIALS NEEDED

- Flower pots and other large plant containers
- Potting soil
- Sand or small stones
- Plants started in preceding activities
- Water

PROCEDURES

1. Choose a plant you have started in one of the preceding activities.
2. Choose a large pot or other container and cover the bottom with small rocks or sand 1 cm. (¼ in.-½ in.) deep.
3. Add potting soil until the container is three-fourths full (don't pack it down).

FIGURE 21-1. Plant being transplanted.

4. Make a hole in the center of the potting soil approximately 3 cm. (1¼ in.) across and 2 cm. (¾ in.) deep.
5. Remove your plant from its present container and put its roots in the hole you prepared in the potting soil. You might need to make the hole bigger if you have many roots.

6. Add potting soil until the roots and any other parts are covered, leaving only the stem and leaves exposed.
7. Carefully water your plant and put it in a sunny place to grow.

TEACHER INFORMATION

Plants with strong, hardy root systems will work best. Transplanting should be done quickly so the roots will not be exposed to the air any longer than necessary.

Transplanting fails most often because of damaged root systems and overwatering.

Ask the children to bring pots and containers from home. An interesting variety of plant containers is likely to result.

ACTIVITY 22: What Are the Parts of Some Common Plants?

MATERIALS NEEDED

- Large paper or large plastic bags
- Spoons for digging
- White paper
- Pictures of plants
- Drawing paper
- Crayons

PROCEDURE

1 Take a field trip to a vacant lot near your home or school. Choose several plants and dig them up. Make sure you get most of the root system.
2. Put your plants in a bag and bring them back to school. Choose one and spread it out on a piece of white paper.
3. Look at the picture of a plant and find the same parts on your plant.
4. Make a picture of your plant and label all the parts.
5. Compare your picture with the one below. Are any parts missing? If so, can you think of why?

FIGURE 22-1. Plant.

TEACHER INFORMATION

Depending on the season, many of the plants may not have all the parts shown. Save the plants from this activity for Activity 23.

ACTIVITY 23: Why Are Leaves Important?

MATERIALS NEEDED

- Plants gathered in Activity 22
- A picture of a plant similar to the one in Activity 22
- Sheet of paper
- Book

PROCEDURE

1. Choose several of the plants you brought back from your field trip and spread them flat on your desk.
2. Study them carefully and compare their roots, stems, and leaves.
3. In what ways are they alike? In what ways are they different? Can you think of reasons why?

FIGURE 23-1. Leaf.

4. Choose some of your prettiest or most interesting leaves and spread them out on a piece of paper. Put a flat, heavy object, such as a book, on top of them. This is called *pressing*. Wait several days and remove the weight. If you would care to preserve your leaves, ask your teacher for help.
5. Leaves are very important to plants and many forms of life on earth. Do you know why? Ask your teacher to tell you more about leaves.

TEACHER INFORMATION

The shape, color, and texture of leaves can be an interesting study. Pressing, preserving, and displaying leaves in creative ways may add aesthetic dimensions to the activity. An excellent way to preserve leaves is to laminate them in a dry-mount press. They may also be preserved by pressing them with a warm iron between sheets of waxed paper. Leaves are important to many plants because they manufacture food through their "chlorophyll factories." Plants also "breathe" through their leaves. In the daytime (during photosynthesis) they give off oxygen. In darkness their chlorophyll factories shut down but the plants still produce carbon dioxide. During this time they also use oxygen or respire as we do. For additional information about leaves, see your encyclopedia.

ACTIVITY 24: What Are the Parts of a Flower?

MATERIALS NEEDED

- Flowers collected on a field trip or large flowers brought from home or large flowers purchased at a store
- Transparent tape
- Magnifying glasses
- White paper
- Roll of plastic wrap

PROCEDURE

1. Put your flower on a sheet of white paper and examine it carefully.
2. Compare your flower to the one in the picture. Can you find the same parts? You may need a magnifying glass to help you.
3. Carefully take your plant apart. First find the petals, then the sepals, and then the pistil. Next find the stamens and the anther.
4. Use transparent tape to tape the parts to your sheet of white paper.
5. Label the parts, cover your paper with plastic wrap, and hang it on a wall of your classroom.
6. Ask your teacher to help you find out why each part of a flower is important.

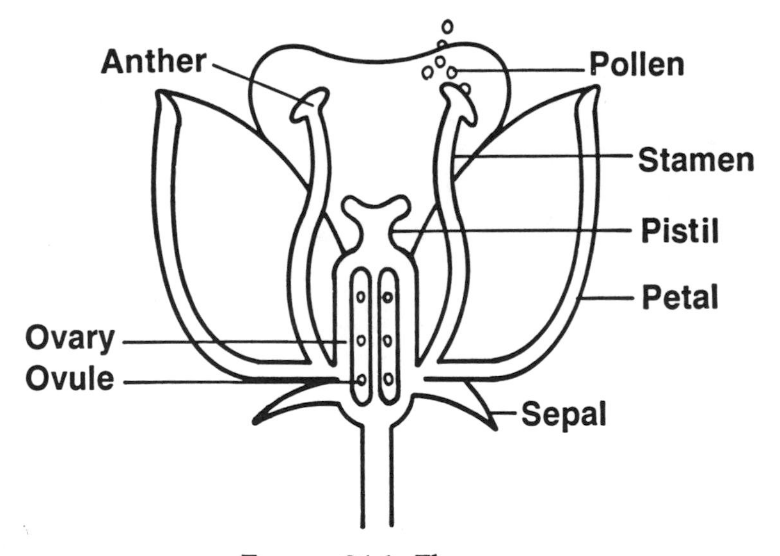

FIGURE 24-1. Flower.

TEACHER INFORMATION

Typical flowers have four sets of parts: sepals; petals; stamens; and the pistil, which includes the ovary and the stigma. Pollen comes from the top of the stamens, which are supported by thin filaments. When the ovary ripens, the top of the pistel, called the stigma, will become sticky to collect pollen spread in the air or by insects. When a flower has all four parts it is said to be complete. Flowers lacking any of the four parts are generally classified as incomplete.

This is an excellent individual or small-group activity. Extension and enrichment activities could include collecting flowers and pictures for bulletin boards, and studying flower arranging. This activity could motivate children to learn the purpose and function of flowers in nature. Flowers should be appreciated for their beauty. Memorizing the names of flower parts is of little value in the elementary school, but careful observation and first-hand experience will increase children's awareness of plants and enhance appreciation for their beauty.

ACTIVITY 25: How Can a Plant A "Maze" You?

(Teacher-supervised activity)

MATERIALS NEEDED

- Utility knife
- Water
- Large shoe box
- Heavy cardboard
- Masking tape
- Small plant such as wandering Jew (genus *Zebrina)*

PROCEDURE

1. Cut a hole in one end of the shoe box 6 cm. tall (2½ in.) and 2 cm. (¾ in.) wide.
2. Use the end of the box as a pattern to cut four dividers (pieces of cardboard) as tall as the box but 2 cm. (¾ in. to 1 in.) shorter than its width.
3. Tape the cardboard dividers upright along the inside of the box, alternating from side to side. Be sure the first divider is attached to the same side as the slot you cut in the box.
4. Put your small plant in the end of the box opposite the slot. A box arranged this way is called a *maze.*
5. Put the lid on the box and *turn the opening toward bright sunlight.* Every three or four days, remove the lid enough to water your plant and observe its condition.
6. Observe your plant for six weeks to two months. What is happening? Can you think of reasons why?

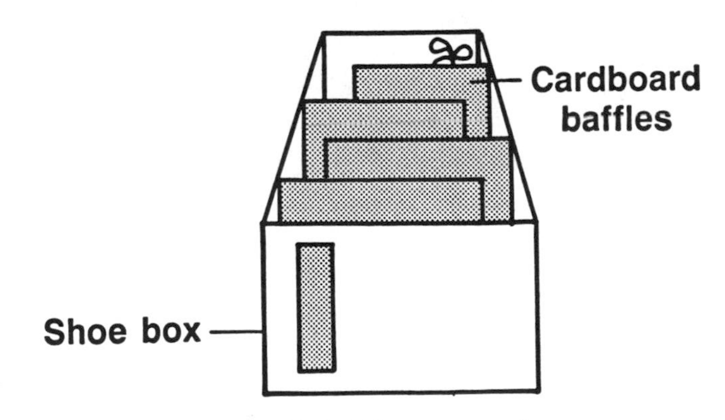

FIGURE 25-1. Plant maze box.

TEACHER INFORMATION

This activity is designed to help children discover that green plants need sunlight and some will travel to find it. After several weeks the plant will grow around the dividers toward the light source. The attraction of plants to light is called *phototropism.* A sprouting potato may be substituted for the wandering Jew, and will not need water.

If you can locate a field of sunflowers near your school, visit it several times on a sunny day. The children should discover that most sunflowers turn their flowers toward the sun and will follow it during the day. (Remember, there are usually some that hear Thoreau's "different drummer" and don't conform.)

ACTIVITY 26: What Is This?

MATERIALS NEEDED

- Bread, jelly, orange, cheese, and other foods
- Magnifying glasses
- Plastic margarine containers and lids

PROCEDURE

1. Choose four different foods and place each one in a container. Put the lids on and place each container in a warm, dark place. (If you choose bread, be sure it is moist.)
2. In four or five days remove the lid of each container and observe the contents.
3. What has happened to the food?

TEACHER INFORMATION

This activity will introduce mold in a controlled environment. Most children have seen mold but only in the context of something that has "spoiled" or been ruined. The following activities will help children learn more about mold and how it is both harmful and helpful in their lives. The containers used in these activities should be clean and thoroughly rinsed. Soap residue may retard the growth of mold.

Most activities are excellent individual or small-group activities. Teacher demonstration and classroom discussion should come *only* after each child has had first-hand experience with the activities.

ACTIVITY 27: What Is This Strange Plant?

MATERIALS NEEDED

- Containers and foods used in Activity 26
- Magnifying glasses
- Drawing paper
- Markers

PROCEDURE

1. Examine the contents of each of your containers.
2. Use a magnifying glass to study the growths on the food in each container.
3. Make a picture of everything you observe.
4. Touch your strange plant. Wash your hands after you touch it. Sniff it without getting too close. Do you notice an odor?
5. Under your picture, write some words to tell what you observed, felt, and smelled.
6. Be prepared to bring your containers and picture to share with the rest of the class.
7. Also be prepared with questions to ask your teacher.

TEACHER INFORMATION

Your media center or school library may have children's books about mold. Children should be encouraged to use the books, but not before they have completed the activities. Elementary and junior high school books are excellent sources for teacher background information and pupil reference for older children. It is recommended that reports be assigned only on a voluntary basis.

CAUTION: After children touch the mold, remind them to wash their hands thoroughly before touching any other part of their body, especially eyes. Children with bronchial problems should be cautioned not to get close to the mold or sniff it. This might be an opportunity to discuss the dangers of tasting, smelling, or touching unfamiliar plants or animals.

ACTIVITY 28: What Does Your Strange Plant Grow On?

MATERIALS NEEDED

- Miscellaneous living (or once-living) and nonliving things, such as cut-up fruit, melons, potato, cheese, bread, wool, nails, magnets, rocks, and wood
- Plastic margarine containers and lids

PROCEDURE

1. Choose five different items from the table.
2. Put each item in a plastic container with the top sealed.
3. On which items do you think the strange plants will grow? On which do you think they won't? Be prepared to explain your reasons.
4. Place your containers in a warm, dark place.
5. After five or six days open your containers and observe the results.
6. Compare your observations with the ideas you had about which would and would not grow. Were your ideas correct? Can you think of why?

TEACHER INFORMATION

This activity is designed to help students see relationships, to reason, and to hypothesize. It has been used most successfully with students well into the concrete operational years, grades four through eight. The most obvious conclusion should be that mold grows on living (or once-living) things and not on nonliving things. Molds use the once-living materials for food. Given enough time and proper conditions, mold will cause wood to rot (see Activity 29), but probably not within the time allowed for this activity.

SUGGESTIONS FOR EXTENSION AND ENRICHMENT

1. We have been growing mold under the controlled conditions of moist, recently living material in a sealed container in a warm, dark place.
2. Select one material such as bread and repeat the activity, changing one variable, for example, dry bread in a warm, dark place compared with moist bread for the same length of time in a warm, dark place. Will dry bread support the growth of mold as well as moist bread? You might try a freezer, which is cold and dark.

ACTIVITY 29: What Have We Learned About Our Strange Plants?

(Total-group activity)

MATERIALS NEEDED

- Containers of mold from previous activities
- Student drawings
- Books and pictures from media center, public library, and home

PROCEDURE

1. Participate in a class discussion. Share and compare your findings with those of the other students.
2. Ask any questions you may have.
3. Use your picture and containers to design a display and bulletin board about your mold.

TEACHER INFORMATION

After studies about mold, teachers often prepare an informational audio tape to be used *after* a class discussion. The tape is optional but it may help answer or reinforce concepts identified during the discussion. Try to help children discover the answers through sharing and reference sources. Avoid telling them more than is necessary. The following are concepts you may want to include on the audio tape or in your summary:

1. Most molds look like cotton. Many are not white but they have a "cottony" texture.
2. Molds grow best in warm, damp, dark places. Mold is a problem in parts of the United States where the climate is humid and warm.
3. Mold often damages food, leather, clothing, and paper. Some molds cause diseases in man, plants, food, crops, and animals.
4. Many molds are helpful. They cause wood, leaves, and other materials to rot, forming humus, which makes the soil rich. Man uses mold to make drugs, such as penicillin. Molds also produce carbon dioxide, which green plants use to make food. Mold grows on Roquefort cheese and helps ripen it.
5. Mold reproduces by releasing spores, which travel through the air or are carried by animals.

ACTIVITY 30: What Are Some Interesting Homes for Plants?

MATERIALS NEEDED

- Potting soil
- Small plants
- Wild bird seed
- Various containers (one-gallon clear, wide-mouthed bottle; clear plastic shoe box; one-gallon plastic milk bottle cut in half; fish bowl)

PROCEDURE

1. One kind of artificial home for plants is called a terrarium. Terrariums are different from other plant homes, such as bowls or pots, because they are often sealed so no air can get in or out, and larger ones may contain both plants and animals.
2. Choose a container and make a terrarium of your own. Use the drawings in Figure 30-1 to help you.

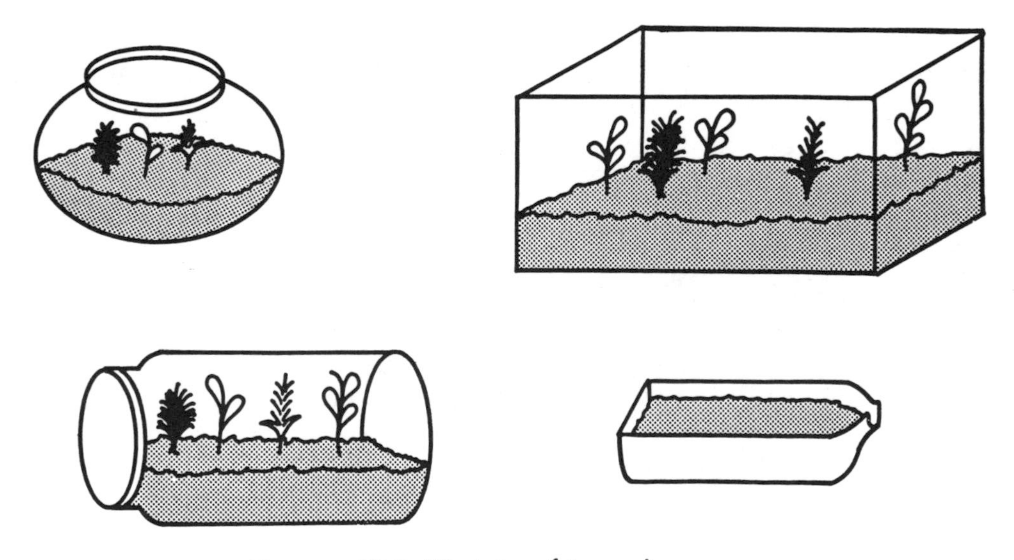

FIGURE 30-1. Variety of terrariums.

TEACHER INFORMATION

Larger terrariums may be sealed and continue to grow and develop for many months. The plants will continue to produce oxygen and carbon dioxide, and moisture will be released in the air and may form water droplets inside the container. The terrarium is then said to be *balanced.* As a class project, you may want to convert a 10-gallon aquarium to a terrarium and include plants and animals such as newts and salamanders. Be sure to check reference sources on terrariums before planning a classroom project.

It is important in balancing a terrarium that you choose plants that require similar amounts of moisture and sunlight. Terrariums in unique containers can make attractive gifts or centerpieces. A plastic two-liter soft-drink bottle with tiny plants and ground cover is an example.

ACTIVITY 31: How Do Seeds Travel?

(Teacher-supervised activity)

MATERIALS NEEDED

- Paper or plastic bags
- Scissors or knives
- Magnifying glasses
- White paper

PROCEDURE

1. Visit a vacant lot or field in late spring or early fall. Be sure you wear long pants and stockings.
2. Explore the field. Try to find plants that have bloomed and are producing seeds or "turning to seed." Walk through the field and examine your pants and socks.
3. Look at trees in your neighborhood (especially in the fall). See if you can find seeds or nuts on or around these trees. Be sure to include a dandelion.
4. From your field trip to the vacant lot and examination of trees, collect as many different kinds of seeds as you can. Be sure to examine your pant legs and socks.
5. Put the seeds you have gathered on a clean, white sheet of paper on your desk. Examine them with a magnifying glass.
6. Seeds are spread or dispersed in many ways. Can you imagine how your seeds might travel?
7. Find a fluffy white dandelion top. Examine one of the tiny, white tufts. Can you find the seed? What does the seed have to help it travel? What makes it travel? Can you find other seeds that might be spread in the same way?
8. Did you examine your pant legs and socks after walking through the weeds? If a dog or other furry animal walked through the field, what might happen?
9. Some plants have special ways of spreading seeds. Ask your teacher about them.

TEACHER INFORMATION

Your library or media center will have books on seed travel. Try to review as many as possible and have them as references for children. Seeds travel most frequently by wind, gravity, and animals.

ACTIVITY 32: Can We Communicate with Plants?

(This is a "just-for-fun" activity)

MATERIALS NEEDED

- Six to eight house plants of the same variety, age, and vigor
- Sensitive scale to weigh grams and ounces (food scale)
- 9″ × 12″ newsprint
- Markers
- Player and records or tapes of different kinds of music
- Name tags

INFORMATION

Scientists have long wondered if there might be a way to communicate with plants. Although there has never been any scientific evidence that plants can talk or hear, you may be communicating with them in special ways through the care you give them. Most plants need sunlight, water, warmth, and the correct kind of soil. Do they need anything else? Just for fun, try some of these things. As you begin, you will probably think of many more ideas to try.

PROCEDURE

1. Make pictures and describe each plant. Measure them. Weigh them and record.
2. Name all but one of your plants and attach the names to their containers. When you come near them to water or move them always use their names and ask them how they are feeling. Care for the plant without a name in exactly the same way but don't talk to it.
3. Put two of your plants with names closer together. Isolate the plant without a name.
4. Play soft soothing music to two of your plants daily, for one hour. Play loud harsh music to two others.
5. Try other ideas on different plants, but remember to do just one thing different to any plant. Keep a record of what you did to each plant. Be sure all plants are adequately watered.
6. After two or three weeks, bring all your plants together. Observe them. Make pictures of them. Weigh them and measure them again. What can you say abut this experience?
7. *Remember,* this is not a scientific experiment. *It's just for fun.*

TEACHER INFORMATION

Although there is no scientific evidence that plants communicate with humans or each other in a linguistic sense, there is some evidence to suggest that certain varieties of plants (trees) may send out a message to others of the same variety when they are under attack by insects or fungus.

People who love plants may ascribe personalities to them, talk to them, and give them better care, on which they will thrive. This could lead to an interesting discussion of how plants should be treated in our society.

ACTIVITY 33: How Can You Make a "Dandy" Tasting Salad?

(Teacher-supervised activity)

MATERIALS NEEDED

- Young dandelion leaves, gathered in early spring before they blossom
- Salt water
- Salt
- Pepper
- Chives
- Salad dressing
- Margarine
- Eating utensils and plates
- Plastic bag
- Heat source
- Saucepan

PROCEDURE

1. Gather a plastic bag full of young dandelion leaves to bring to class.
2. Pull the young plants up, being careful to get as much of the root as possible.
3. Pull the leaves from the root, rinse them and boil in salt water for about 10 minutes. Season the cooked leaves with salt and pepper and a pat of margarine. Eat your special vegetable.
4. The crowns (tops) of the roots may also be prepared in the same way to provide another variety of vegetable.
5. Mix young dandelion leaves with lettuce and celery leaves at about the ratio of one dandelion leaf to two lettuce and celery leaves. Add chives and mix thoroughly.
6. Use salad dressing to taste.
7. You now have a new special inexpensive salad to show at home.

TEACHER INFORMATION

Dandelions *(Taraxacum officinale)* must be picked while they are very young. Boiling in salt water will help remove the bitter taste. *Be sure the leaves are not gathered in areas that have been sprayed.* The bulbs of roots may be treated the same as leaves. Even young dandelion leaves may be somewhat bitter. In using them in fresh salad, be sure to use plenty of lettuce. This may be an appropriate time to remind children not to eat unknown leaves or other parts of plants, such as berries or flowers. Dandelion leaves are a source of vitamins A and C.

ACTIVITY 34: What Parts of Plants Do You Like?

MATERIALS NEEDED

- Assorted fresh fruits and vegetables cut in pieces
- Nuts
- Raisins
- Grains
- Soybeans
- All-grain breakfast cereals
- Margarine
- Bread or rolls
- Various juices in small paper cups
- Copy of the classification activity sheet for each student
- Pencils

PROCEDURE

1. Select ten pieces of plant parts and juices from the table and take them back to your desk.
2. Try to organize and classify the foods using the "Fruit and Vegetable Classification Chart."

TEACHER INFORMATION

This activity should help children understand the value of plant parts as a food source. This can be an opportunity to help children become acquainted with a wide variety of food plants. Save as many plants as possible for the next activity. Children can work individually or in small groups to prepare a basic list and then come together to develop a master list and discuss all plant foods in the room.

Name ______________________ Date ______________________

FRUIT AND VEGETABLE CLASSIFICATION CHART

Common name of plant	*Fruit or vegetable?*	*Part of the plant we eat*	*Cooked, raw, or both?*	*My favorite*

ACTIVITY 35: What Have We Learned?

(Teacher-supervised activity)

MATERIALS NEEDED

- Assorted fresh fruits
- Vegetables
- Nuts
- Juices
- Salad greens
- Sprouts
- Bread or rolls and margarine from previous activities
- Information and classification lists developed in previous activities
- Paper plates
- Cooking utensils
- Bowls
- Plastic knives, forks, spoons
- Seasoning
- Salt
- Pepper
- Napkins
- Stove or hot plate

PROCEDURE

1. Use the list of plant foods to plan a luncheon meal.
2. Make a menu describing the items you will prepare and serve. Try to use as many adjectives as you can to make everyone want to try all the foods you prepare.
3. Have your teacher help cut and cook your meal.
4. Invite special people to have lunch with you. Don't forget the school cook or cafeteria workers.

TEACHER INFORMATION

This activity combined with the dandelion salad activity can be an enjoyable culmination to this study. In addition to the food preparation and serving, students may choose to share reports, pictures, songs, and other information acquired during the study. Planning and writing the menu, designing invitations, giving reports, sharing art work, singing songs, decorating the room with plants and flowers they have grown, and learning how to be good hosts and hostesses should be a planned part of this activity.

Note: You will need to plan well in advance to arrange for parents and aides to assist with the preparation and cooking. A prior activity on bread making could be developed (good place for a mother or father to be used).

Section 3

ANIMALS

TO THE TEACHER

The study of animals is so broad that it can and often does encompass a lifetime. In this section we have tried to limit our study to some of the animals in our immediate environment. We have focused on the following general areas:

- Pets and Imaginary Animals
- Relatives of Pets (Animal Tracks)
- Animals in Our Neighborhood (Nature Square)
- Animals outside Our Window (Birds)
- Animals in Our Room (Aquariums; Terrariums)

Many broad categories such as domesticated animals, animal colonies, animals of the zoo, and exotic animals have been omitted. You are encouraged to study these areas, using the skills and resources developed in the following activities.

This section places a strong emphasis on the use of books, other media, and resource persons. Before you begin, read through the materials lists so you can begin organizing, collecting, and arranging for future activities.

The nature square requires outside activities and field trips.

There are also several simple construction projects such as birdfeeders, birdhouses, and butterfly nets. Enlist the help of parents and others.

To complete some activities, small costs may be involved. Inexpensive hand lenses, animals, and plants may need to be purchased. In some cases, orders may have to be sent to science supply houses and delivery time will be a factor.

We urge you to use resource people whenever possible. There are amateur as well as professional naturalists everywhere.

The main purpose of these activities is to help students develop respect and appreciation for the variety, beauty, and wonder of the animals around them.

Be as creative as you can. Wander and explore in any direction you choose. These activities could be starting points for lifelong interests and hobbies.

All of the following activities can be adapted for use with young children. Occasionally, specific suggestions for adaptation are made, but usually we have attempted to present ideas that are flexible enough to be used with any age group.

Perhaps to prepare yourself for this adventure you could read Robert Frost's poem, "The Vantage Point." You can find a vantage point wherever you are.

ACTIVITY 36: Who Are the New Members of Our Class?

(Total-class activity)

MATERIALS NEEDED

- Two or more animals of different kinds, properly caged for a two- or three-week stay in the classroom (birds, fish, turtles, frogs, gerbils, or guinea pigs are examples)
- Pencils
- Copy of "Animal Study Sheet" for each student

PROCEDURE

1. Today we have several animals in our room. Observe each animal carefully. Make a list of everything you can about each animal.
2. What kind of animal is it? How many legs does it have? What is its outer covering like? What kind of a mouth does it have? How does it move about? Can you tell what it eats?
3. Your teacher will give you a paper with some questions on it. Observe each animal and see how many of the questions you can answer. Save your paper so you can report your findings to the class.

TEACHER INFORMATION

This introductory activity is planned to help children begin to develop a systematic awareness in their observations of animals. From the group discussion, children should be able to identify similarities and differences in some kinds of animal life. An outcome from the group discussion should be the planning of care for the animals while they are in the room (see Activity 37).

It is *not* recommended that strange, unusual, or possibly dangerous animals be introduced at this time. Snakes, spiders, and wild animals should be reserved for a later time in the study.

The "Animal Study Sheet" shown here can be reproduced in quantity and used throughout the time animals are being studied. You can modify this sheet according to the ages of the children, geographical regions of the country, and animals you choose to study. In early childhood education, a simplified version of this sheet can be developed in chart form for large group use.

Before beginning a study of animals be certain to check with your library and/or media center for books, stories, filmstrips, pictures, and so forth. Many books and pictures about the specific animals you introduce should be available in the classroom.

Name ______________________ Date ______________________

ANIMAL STUDY SHEET

Name of animal ____

Type of animal ____

Size of animal in centimeters or inches ______________________

Color(s) of animal ____

____

Outer covering of animal _____________________________________

____

Locomotion of animal (How does it move?) _____________________

____

Number and kind of appendages (legs, arms, wings, fins, tail) of animal _____

____

Type of food(s) eaten by animal _______________________________

____

Habitat of animal (Where does it usually live?) _______________

____

Tell as many special things as you can about this animal ______

____

____

____

____

____

____

____

____

____

ACTIVITY 37: How Can You Keep Your Pets Healthy?

MATERIALS NEEDED

- 24″ × 36″ chart paper
- Marking pen
- Completed "Animal Study Sheets" from Activity 36

PROCEDURE

1. You now know the kinds of pets with which you are going to share your classroom. Your "Animal Study Sheet," which you filled out for each pet should help you decide what your pets need in order to be well and comfortable during their stay with you.
2. With the help of your teacher see if you can find answers to the following questions:
   a. What kind of housing is best for each pet?
   b. What kind of food does each pet need? How much? How often?
   c. What kind of environment does each pet need or like best (water, dirt, rocks, grass, perches)?
   d. Can more than one pet live in the same area?
   e. What temperature does your pet need or like best?
   f. All living things must get water in some way. How does your pet get its water?
   g. What are other special needs you think your pet has?
3. Make a chart for each pet, answering the questions listed above.
4. Divide the class into groups, and, using the chart as a guide, care for each pet for a week.

TEACHER INFORMATION

Many of the activities in this section begin with total-group planning or discussion, but in each case they should lead to individual participation. Through caring for classroom pets, children should become acquainted with the needs of specific animals and be able to generalize their experience to the needs of many animals. This activity is intended to introduce a major concept—*responsibility*—which will be emphasized throughout the study. Animal ownership implies the responsibility for the care, welfare, and behavior (control) of the pet. Later, the Big Idea of responsibility will be extended to include all kinds of wildlife in our nation and the world.

ACTIVITY 38: How Many Pets Can You Find?

(Total-class activity)

MATERIALS NEEDED

- Paper
- Pencils

PROCEDURE

1. Many people have pets. A pet can be any animal that lives around your home and that you care for. The most common pets are dogs, cats, birds, and fish but many people have other unusual pets.
2. With your group, try to make a list of all the kinds of animals people might have for pets.
3. Decide which of the pets on your list would be practical to keep in your area. Draw a line under those animals and be prepared to report to the whole class. Choose someone from your group to report.
4. Your group leader should report to the class the animals you have chosen. Your teacher will list them on the board.
5. How many of the pets listed on the board does someone in your class have? Try to plan a time when each of these pets could come to school for a short visit.

TEACHER INFORMATION

We suggest you divide the class into groups of five or six to do steps 1-3. Then have group leaders report to the whole class.

This activity should be conducted early in the study so that pets can be scheduled throughout the time you are working with animals. Care should be taken to ensure that the animals are not dangerous and are properly housed or controlled by a restraint. A responsible individual should accompany each animal. Prepare the class for the animal visitor so they won't frighten or disturb it.

Activity 39 will suggest other ways to locate unusual animals.

ACTIVITY 39: How Many Other Pets Can You Find?

MATERIALS NEEDED

- Paper
- Pencils
- Calendar
- 9″ × 12″ drawing paper
- Crayons
- Record player
- Camera

PROCEDURE

1. Now that you have found out how many kinds of pets people in your class have, you may be surprised at how many more you can find.
2. Choose a group of students to make a survey of your entire school. Make a form that can be passed out (with permission of your principal and other teachers) to all students in your school. Ask if they have a pet or know of anyone else, not a student at your school, who does. You might mention that you are interested in unusual pets. Be sure to get the names, addresses, and phone numbers of the people who own the pets.
3. When the questionnaires are returned, develop a system for classifying or grouping them.
4. As a class, make a list of all the pets people in your class own, plus the ones recommended in your questionnaires.
5. Look at a calendar and estimate the number of days you are planning to spend studying animals. From the list developed in step 4, choose the pets you would most like to have visit your class. You may be able to have more than one a day if you choose carefully. Also, if you find many unusual or interesting pets, they could be scheduled for visits throughout the year. Mother pets with babies will need to come when their owner feels it is best.
6. Contact the owners and arrange to have the pet come on a specific day. If possible, ask the pet's owner to be prepared to tell the class about the pet.
7. Before your "guest" pet comes and while it is there, use your library and other resources to find out as much as you can about it. Record all the information you can. Make drawings and pictures. Invite other classes to visit and tell them about your pet. Make a photograph of each pet. If your guest pet makes sounds, record them.
8. Make a scrapbook of your guest pets for the year.
9. Be sure to write thank-you notes to people who share their pets with you.
10. Which pets would you most like to have? Why?

TEACHER INFORMATION

This activity requires careful coordination and planning. Be sure to consult the principal of your school in the early stages of planning. If more than one class will be studying animals during the year, be sure to involve those teachers in the planning.

Don't forget to contact pet store owners and veterinarians if they are available in your area.

If you live near a zoo, docent services are often available to bring unusual animals to the classroom. It is recommended that exotic animals be scheduled late in the study, after the children become comfortable with animal visitors.

Adequate housing and control of the animals are essential. Children should understand that animals become frightened easily in strange surroundings.

You may be surprised at the pets you find.

ACTIVITY 40: What Kind of Pet Would You Like to Own?

MATERIALS NEEDED

- 9″ × 12″ drawing paper
- Pencils
- Crayons

PROCEDURE

1. From the list of all the pets you could have, choose one you would like to own.
2. Make a picture of your pet. Draw a picture of the place your pet will need to live.
3. Write about, or draw a picture of, the food and water your pet will need each day and how you will provide it.
4. Write in one column the advantages (good things) about owning the pet you have chosen.
5. In another column list the disadvantages (problems) of keeping your pet.
6. Show your picture and read your lists to a group of students or the whole class. Perhaps they can help you decide whether this is a pet you should have.

TEACHER INFORMATION

People occasionally acquire pets they cannot care for properly. This activity attempts to introduce the idea of responsible pet ownership. If you live in an incorporated area, there may be laws regulating the ownership and care of animals. Animal ownership may also be regulated because of health factors (disease transmission).

Wild animals do not make good pets. This concept will be developed later but can be introduced at this time.

ACTIVITY 41: What Can Hair Tell You?

MATERIALS NEEDED

- Large, gentle dog
- Newspapers
- Markers
- Clean, wire pet brush
- Small plastic bags
- Small labels

PROCEDURE

1. Have your teacher help you find a large, very gentle, well-trained dog with a shaggy coat. A dog who lives outside most of the year is best. Do *not* use a stray.
2. Spread newspapers on the floor. Use a clean, wire pet brush to brush your dog ten times (strokes). Remember the exact area you brushed. You might want to make a drawing of the dog to help you remember.
3. Collect all the dog hair from the newspaper and wire brush. Store the hair in a small plastic bag. Put a tag on the bag telling the day, date, and time of the brushing.
4. Brush your dog in exactly the same place, in the same way, four times during the school year. September, December, March, and late May might be best.
5. Examine and compare the hair you collected after each brushing.
6. What can you say about this?

TEACHER INFORMATION

Before you brush the dog, it might be well to have it visit several times to get acquainted with the class. This activity is most effective in climates with pronounced seasonal changes. However, most dogs will undergo seasonal changes in the thickness of their coats, even in mild climates.

Animals with hair shed constantly (humans included). Animals who normally live outdoors often develop a thick undercoating in winter. The condition of the hair of a mammal is often an indicator of the animal's health (and of humans').

Students might notice that they lose hair regularly. Unless the loss of hair is extreme or in patches, it is part of the normal hair growth pattern. At any one time about 29 percent of human hair is in the "resting" stage; that is, the hair is gone but the follicle will remain and in a few months will grow a new hair. Baldness, which occurs mostly in males, is hereditary and so far is irreversible, though remedies are sought.

The fur of many mammals serves the dual purpose of warmth and protection. Human hair mostly provides protection. Hair on our head insulates somewhat, but mostly protects the skull. Eyebrows shade our eyes from the sun.

Eyelashes and hair in the nose and ears keep dust and insects out. Some scientists believe the tiny sensitive hairs around our mouths serve the same purpose as feelers in insects.

All hair is not the same shape. If you examine different types of hair under a low-power microscope or magnifying glass you will find that straight hair is round. Curly hair is flat.

This activity is also in "Starter Ideas."

ACTIVITY 42: How Many Play Animals Can You Find?

(Lower-grades activity)

MATERIALS NEEDED

- Stuffed animals
- Animals made of other nonliving materials

PROCEDURE

1. Many of us have animals in our homes that are not living. Some of these animals are used for decorations, some just for fun or play. Do you have a stuffed animal or some other kind of nonliving "friend" you play with?
2. With your teacher, choose a day when all of these nonliving friends can come to school.
3. Introduce your friends to the class and tell about them. What are their names? Where did they come from? Why are they special?
4. Think of as many ways as possible to tell a living from a nonliving animal.

TEACHER INFORMATION

One purpose of this activity is to encourage language experience through sharing. Another purpose is to introduce the idea of living and nonliving things. Young children often associate living things with movement or some observable change in behavior. Life processes, needs, and changes are best understood if developed through concrete experiences and observations.

Science today is studying the basic building blocks or substances of life. There is much debate as to what constitutes living and nonliving material and what the necessary characteristics of life really are.

For preoperational and concrete operational children, generalizations about basic needs of life such as food, water, and reproduction seem to be sufficient for a study of the life sciences. Specific characteristics of plants and animals are introduced when appropriate.

ACTIVITY 43: How Can You Make a New Animal for a Funny Zoo?

MATERIALS NEEDED

- 9″ × 12″ art paper
- Watercolors
- Paintbrushes
- Crayons
- Water

PROCEDURE

1. Use your paintbrush to get your paper wet.
2. Put three or four watercolors on your paper. Let them run together.
3. After your paper is dry, look at it carefully. Try to imagine there are strange animals on your paper. How many can you find? What else do you see?
4. Use your crayons to outline (draw around) the animals and other things. Have you ever seen anything like them before?
5. When you have finished, try to think of names for your animals.
6. Find a place in your classroom to hang your funny zoo pictures.

TEACHER INFORMATION

This activity can be adapted in many ways. Large pieces of butcher paper may be substituted with several children working together. Thin poster paint could take the place of watercolor. If the colors are put on the paper in advance, the time required for the activity will be reduced.

In addition to being a form of creative art, this activity should help children to review body parts and animal anatomy informally.

Although many children will only be able to find "real" animals such as birds, fish, etc., in their pictures, they should be encouraged to imagine, explore, and create as much as possible.

Before introducing the activity, you might read or show some children's stories that use imaginary animals, such as the Dr. Seuss stories.

ACTIVITY 44: What Can You Do with Bones?

MATERIALS NEEDED

- Large box of dried, clean bird bones from turkeys, chickens, etc.
- Styrofoam squares 25 cm. × 25 cm. (10″ × 10″) for bases
- Light, soft iron wire
- Strong glue
- Cardboard box
- Masking tape
- Newspapers
- Pictures of dinosaurs and other animal skeletons (including birds)
- Styrofoam balls of various sizes
- Colored plastic clay
- Pipe cleaners
- Wood dowels (miscellaneous lengths)
- Crayons

PROCEDURE

1. Choose several friends and look at the pictures. Could you create and build your own bony animal?
2. Draw a picture of your imaginary animal. Remember, it should have the same basic parts (feet, legs, other parts) as other bony animals do. Label the parts.
3. From your teacher's collection of bones, choose the ones you think you can use to build your animal. If you can't find the right bone perhaps a piece of clay, a pipe cleaner, or a dowel will do.
4. Lay the bones and other parts on a newspaper as if your animal were lying on its side. Remember you may need two or more of some bones if your animal has several arms or legs.
5. Begin to put your animal together from the bottom up. Use a Styrofoam square as a base. Build the feet first, then the legs. Use glue and iron wire to hold the bones together. You may need to wrap the joints in masking tape until they dry.
6. If your animal has only two feet, you may need to use something to hold it up as you add the backbone and upper body parts.
7. If you have trouble finding bones for the head, consider using Styrofoam balls or clay.
8. When your animal is finished, you may need to use dowels or coat-hanger wire to help it stand. Be sure to wire or glue its feet to the Styrofoam base.
9. Think of a name for your animal.
10. Write information about your animal on a piece of paper. Tell what it ate, where it lived, and how it moved about. Be sure to include a picture showing how the animal looked when it was alive.
11. Invite other people to look at your museum.

TEACHER INFORMATION

The purpose of this project is to determine how well children understand the bony skeletons of animals, to encourage creativity, and to give practice in constructing three-dimensional figures from two-dimensional pictures or plans. Encourage children not to reconstruct turkeys or chickens. Instead, ask them to use the bones to build what they think would be an "ideal" animal.

Before children begin the "Procedure" steps, use a picture to review the names and basic functions of a typical skeleton in general terms; that is, feet, legs, hips, backbones, ribs, neck, and head.

This activity could be done any time, but after Thanksgiving would, of course, be best. Bird bones are recommended because of their light weight and because they can be easily cleaned and dried. Before Thanksgiving send a note to parents asking them to contribute bones. Before they are sent to school, the bones should be stripped of flesh and then dried thoroughly in a hot oven. Check the bones the children bring to be sure they have been prepared properly, otherwise odors may develop. If children want to add wings, tails, eyes, or other nonbony parts, suggest colored paper.

If you teach young children, you may want to simplify the construction by using small bones, pipe cleaners, and Styrofoam balls.

ACTIVITY 45: What Can Fooprints Tell Us?

MATERIALS NEEDED

- Dark-colored fingerpaint
- Newsprint

PROCEDURE

1. Clench one hand into a fist. Use the fingers of your other hand to lightly coat the bottom side of your fist (opposite the thumb) and the outside of your little finger with paint.
2. Press the painted side of your fist on the paper.
3. Open your fist and put a light coat of paint on the tips of each finger (not the thumb) of the same hand.
4. Gently press the painted fingers about 1 cm. (½ in.) above the widest part of the impression you made with your fist.
5. What does the print look like?

FIGURE 45-1. Baby's footprint.

TEACHER INFORMATION

The impression should resemble a baby's footprint.

Newborn babies are sometimes footprinted for positive identification. Most students may have heard of using fingerprints for identification, as each person has a unique set. Footprints serve almost as well and are used with babies because their fingers are so tiny. Point out to the children that although each human footprint has unique swirls, they all have the same basic shape.

ACTIVITY 46: What More Can Footprints Tell Us?

MATERIALS NEEDED

- Photograph of Neil Armstrong's footprints on the moon (see the December 1969 issue of *National Geographic*)
- Diagram of bird tracks (see Figure 46-1)
- Crayons
- 9″ × 12″ sheet of newsprint for each student

PROCEDURE

1. Look at the picture. It is a print of the most famous animal track of this century. Can you guess what it is?
2. Look at the footprint you made with your hand in the preceding activity. Watch your feet as you take several steps. On a piece of paper draw a picture of what your bare footprints would look like if you walked four steps in sand or mud.
3. The tracks of animals can tell us many things. Study the picture that shows the tracks of three birds standing and walking in the sand. Carefully observe their tracks and see how much you can tell about the birds as they near the edge of a freshwater pond.

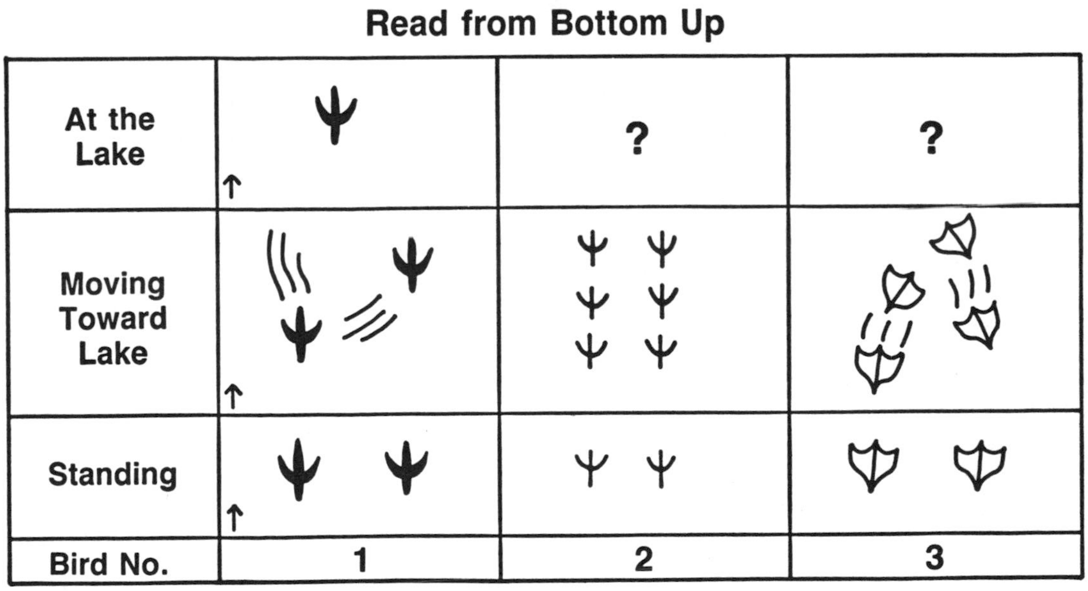

Read from Bottom Up

At the Lake	↑	?	?
Moving Toward Lake	↑		
Standing	↑		
Bird No.	**1**	**2**	**3**

FIGURE 46-1. Diagram of bird tracks.

TEACHER INFORMATION

Your library or media center will have pictures of the astronauts' footprints on the moon. In snow, "moon boots" will make similar imprints.

Before the students do step 2 under "Procedure" you may want to have a student walk and point out the heel-toe rolling motion of our feet inward as we walk. Our feet turn outward. If we walk with our feet turned inward it is called "pigeon-toed." If pigeons are common in your area have students observe them and describe how they walk.

By carefully reading the tracks in Figure 46-1, you may reach some of the following conclusions:

1. Bird 1 is larger than bird 2 or 3 (biggest feet).
2. Bird 3 swims in water (webbed feet).
3. Birds 1 and 3 walk. Bird 2 hops.
4. Bird 1 has longer legs than bird 3 (longer stride).
5. At the edge of the pond, bird 1 is standing on one leg, possibly a crane, heron, or larger shorebird. Bird 3 is swimming away in search of food, probably a duck, small swan, or goose.Bird 2 is probably a perching, hopping, land bird who stopped for a drink and is now perched in a nearby tree.

Activities using animal tracks or signs can help children develop skills of observation and deduction.

ACTIVITY 47: What Other Clues Can Help Us Read Tracks?

MATERIALS NEEDED

- Drawings of common animal tracks of dog, cat, squirrel, deer (numbered but unnamed)
- Paper
- Crayons

PROCEDURE

1. Study the animal tracks in the picture. You have probably seen one or two of these tracks before in dust, dirt, sand, or snow.
2. Compare the tracks. How many toes can you count on each? Which have claws? Can you tell how many have footpads? What is different about track 4?
3. The tracks belong to the following animals: squirrel, dog, cat, and deer. Think of each animal. On your paper write the number of the track above the name of each animal.
4. Compare your answers with others in the class.
5. Animals that belong to the same family make similar tracks. If you changed the size and the shape slightly you could rename the tracks: mountain lion, wolf, elk, chipmunk, groundhog, moose, coyote, or bobcat.
6. Classify the animals named in steps 3 and 5 according to their families.
7. Compare your list with those of others in your class. Ask your teacher for the correct information.

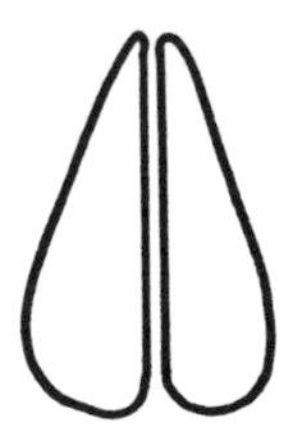

FIGURE 47-1. Comparing four animal tracks.

TEACHER INFORMATION

The drawings in the figure are tracks of a dog (1), cat (2), squirrel (3), and deer (4).

Other members of the same family whose tracks are similar are:

Dog	Cat	Squirrel	Deer
Wolf	Mountain Lion	Chipmunk	Elk
Coyote	Bobcat	Groundhog	Moose

The purpose of this activity is to help children understand that although animal tracks are different, members of the same families make similar-shaped tracks. Thus the track of a dog accompanying a hunter may make a track similar to that of a wolf or a coyote. There may be several varieties of squirrels and chipmunks in the park whose tracks look alike. Tracks give only one clue to the identification and behavior of animals.

For cats, many students will probably choose a track with claws because of the association of cats with being scratched. Members of the cat family, with one exception, have retractable claws that are usually drawn into a sheath when they walk and extended when they climb, attack, or defend. The cheetah, the fastest runner on earth for short distances, is the only cat without retractable claws.

ACTIVITY 48: What Animals Live Around You?

MATERIALS NEEDED

- Paper
- Pencils

PROCEDURE

1. Animals live almost everywhere on earth. There is a place near your home or school where many animals live. You may not have noticed them because they are small or shy, or they live under some kind of cover. Think of your school grounds, backyard, empty lot, park, or even an alley near you. It should be a place that has dirt, rocks, green plants, shade,and some source of moisture. A pond, pool, standing water, or regularly sprinkled area is best. On your paper write the name of the place you have chosen.
2. Describe the area you have chosen as you remember it. Write all the details you remember. Make a picture if you can.
3. Describe your special place to the other class members and your teacher.
4. Make a plan to visit your area and study it. Work alone or choose a friend to join you.

TEACHER INFORMATION

This is the first of a series on the study of a *nature square*. The initial study should be conducted during mild weather in the spring or early fall if your climate has distinct seasons. In the following activities the children will be asked to visit the square regularly, observe the environment, collect samples, and observe changes.

Since the kind of nature squares that are available will be determined by your location (climate; geography; urban, suburban, or rural area), you will need to approve and perhaps suggest locations if the children have difficulty choosing a place.

The major purpose of the nature square activity is to help children become aware of the many forms of animal life that exist around us. Skills of observation, organization, and classification will be emphasized.

If you have a large, landscaped school ground that is several years old, or a large vacant lot adjacent to the school, the entire class could set up nature squares in the same area under your supervision. Although you will have less supervision, there are advantages to having other squares scattered over a wider variety of areas. A compromise might be to choose an area close to the school, where you can help the children learn observation and collection skills that they can apply to the play they have chosen.

Nature square activities are easily adapted for teaching younger children. Choose one or two areas near your school and adapt the following activities and suggestions for use with the total group.

ACTIVITY 49: What Do You Need to Study a Nature Square?

MATERIALS NEEDED

- Meter stick or string 10 meters (10 yards) long
- Wooden stakes or markers to show borders
- "Observation Chart for Nature Square" (one per student)
- 9" × 12" heavy white paper (for mapping)
- Crayons

PROCEDURE

1. Study the list of materials on the Observation Chart. This is the beginning of a list of things you will be looking for when you visit the special place you have chosen, which will be called a "nature square." Notice there are extra spaces at the bottom for you to add unusual or different things you find.
2. When you first visit your square, use the string or meter stick to make approximate measures of its size. Remember your square does not have to be a *real* square shape. It could be long, round, or kidney shaped. Put wooden stakes or markers around the border.
3. On your first visit to the square, make a map-picture of the distinctive features (things that stand out) such as rocks, trees, bushes, or water areas.
4. Report your initial (first) findings to the class.

TEACHER INFORMATION

In the following activities students will be expected to make periodic visits to the places they have selected. Try to screen their choices so their areas will be inviting enough to attract as many forms of wildlife as possible.

Areas should be large enough to include the features mentioned in step 1 of the preceding activity but not so large as to be unmanageable. Shape is important only if it can be altered to include special features such as a tree, pond, or shaded area.

If students or their parents have simple cameras, actual photographs could be included but should not replace the picture-map activity.

Activity 50 gives specific instructions for using the observation chart.

OBSERVATION CHART FOR NATURE SQUARE

Observations, Discoveries, and Conditions

Date of visit				
General conditions: climate, wind, temperature, precipitation, and so on				
Vegetation: trees, shrubs, weeds, flowers, grass, and so on				
Small underground animals: ants, worms, beetles, bugs, and so on				
Small animals on the ground and under rocks, leaves, or other ground cover: spiders, ants, and so on				
Small animals on or in vegetation: aphids, caterpillars, larvae, eggs, bugs, beetles, and so on				
Small flying insects: bees, beetles, gnats, butterflies, mosquitoes, grasshoppers, dragonflies, and so on				
Larger underground animals: gophers, chipmunks, moles, and so on				
Larger animals flying, nesting, perching, walking (tracks), or crawling				
Animals in standing water, pond, or lake: tadpoles, frogs, fish, snakes, salamanders, insects, and so on				
Other observations:				

ACTIVITY 50: How Can a Chart Help Us Make Observations?

MATERIALS NEEDED

- "Observation Chart for Nature Square" (see Activity 49)
- Pencils

PROCEDURE

1. This chart is designed to help you study your nature square.
2. The column on the far left suggests specific things you should watch for and record each time you visit your square.
3. The other smaller columns with boxes are places for you to record words to report each observation.
4. The bottom space is blank all the way across. Use it to record and report things you find that are special about your square.
5. With your teacher and other students, discuss the areas listed in the far left column. Leave out anything that does not apply to your own square.
6. Make visits to your square every other day. Write the date at the top of the column and use the chart as you study your square.

TEACHER INFORMATION

The observation chart is general in nature and may need to be adapted to your particular geographic location.

Discuss each category on the left and be certain students understand the examples. Add and delete items as necessary.

It may be helpful to go through an imaginary visit, step by step, to show them how to use the chart.

Activity 51 suggests guidelines for equipment to take along on visits to the square.

ACTIVITY 51: What Equipment Can Help Us Study Our Nature Square?

MATERIALS NEEDED

- Shoe boxes
- Magnifying glasses (hand lens)
- Metal spoons
- Dull table knives
- Scissors or garden shears
- Tweezers (or forceps)
- Mosquito netting
- Small-to-large jars with lids or assorted rigid plastic containers with lids.
- Small cans with hinged lids
- Small-to-large plastic bags
- Butterfly nets
- Notebook and pencils
- Tape, string, and rubber bands

PROCEDURE

1. Look at the list of materials. You can probably gather most of the equipment around your home and school.
2. Put a mark by the items you think you can find.
3. Discuss the list with your teacher and class. Perhaps you can trade with others to get the things you don't have.
4. After the class discussion make a list of things you still need and ask your teacher for help in locating them.
5. Gather the materials. Put the smaller items in a shoe box and the larger items in a plastic bag (pack glass in paper).
6. Scientists called *naturalists* study nature as you are going to do. The simple equipment you have is very similar to the equipment they use.
7. After you have gathered your equipment, you are ready to begin investigating your nature square just as a naturalist would do.

TEACHER INFORMATION

A simple plan for making butterfly nets is shown in Figure 51-1. Constuction can be done by every child or they can work in groups of three and four.

Before you begin studying and collecting specimens discuss the following simple safety rules:

1. Don't put anything you find in your mouth.
2. Be very careful with insects that sting, such as bees, hornets, and wasps (never disturb their nests).
3. Use a spoon or envelope to scoop crawling insects into containers.
4. If any animals are removed from the nature square, they should have comfortable housing and later be returned alive.
5. Always wash your hands as soon as possible after visiting the nature square.
6. Attend to accidental cuts, scratches, or bites as soon as they occur.

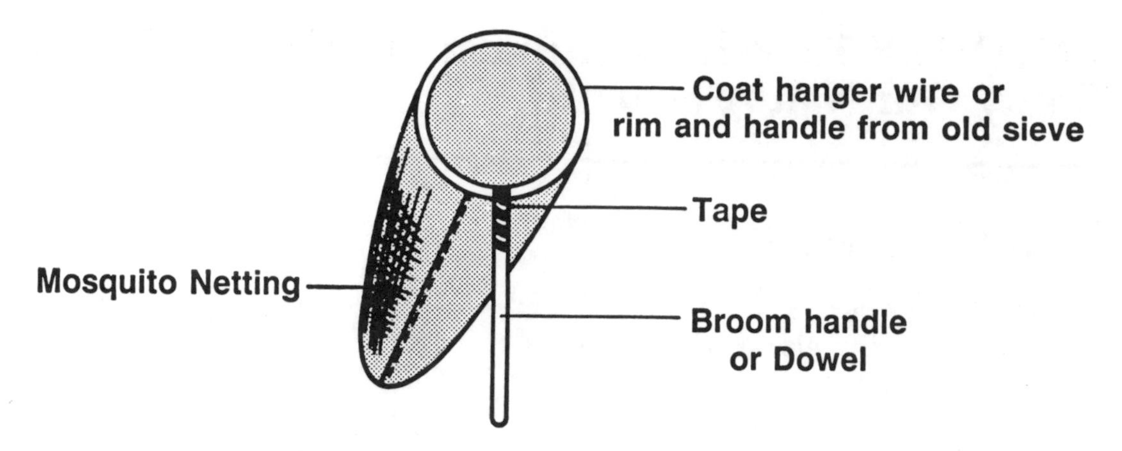

FIGURE 51-1. Diagram of a butterfly net.

If there are poisonous plants or animals common to your region of the country, such as poison ivy, poison oak, ticks, or chiggers be sure the children can identify and avoid these specimens.

Magnifying glasses may be the most difficult item to obtain. High-quality lenses are not necessary for field study. Lenses molded from plastic are sold in the school equipment or stationery departments of most general merchandise (variety) stores. You will, however, need several high-quality magnifying glasses for classroom use.

Items not included in the kit are insect killing jars and materials to preserve and mount specimens. The focus of this section is on the observation and study of nature as it exists. If specimens are brought to school, every effort should be made to put them in a large, suitable container and when appropriate try to include a portion of their immediate surroundings—twigs, leaves, soil. A small, moist sponge or piece of apple, pear, or peach, and plenty of fresh air should be provided for insects. Jars with small holes punched in the top do not provide enough air circulation. Use a piece of mosquito netting held with a rubber band over the top of the container.

After study, all live specimens should be released where they were collected.

Caution students not to bring injured or baby birds to the classroom.

This could be an excellent opportunity to involve parents and other groups. They can help in collecting and organizing equipment.

ACTIVITY 52: How Can We Report Our Findings?

(Class discussion with older students)

MATERIALS NEEDED

- Paper
- Pencil
- Calendar (optional)

PROCEDURE

1. Plan to visit your nature square about every other day for a period of two or three weeks. Use a calendar to write the specific dates you plan to go. When you sign your name to this paper and give it to your teacher it will mean that you agree to visit your square on the days listed, spend at least 30 minutes, make observations, collect specimens when possible, and keep a record on your observation chart. When your teacher accepts your paper it will mean that he or she agrees to assist you in your study and in the evaluation of what you have done. This then becomes a *contract.*
2. Plan with your teacher and other members of the class how your study will be shared. Scientists often hold conferences to read papers to each other. Often they give oral reports from notes. Sometimes they meet in small informal groups to share and compare their important discoveries. Some get together to write books, or magazine and newspaper articles. They usually find some way to communicate their important findings to others. As a group of naturalists, how will you report your findings so others will know what you have discovered?

TEACHER INFORMATION

Omit step 1 if you have made the study during class time on the school grounds or a nearby vacant lot. *Contracting* is a technique some teachers have found helpful in encouraging responsibility and independent work. Parents are often asked to sign a contract too.

The length of time students spend studying their squares will vary according to the situation, but through experience we have found that stating a specific minimum time is helpful.

Students may need help in developing skills of observation. Remind them to move slowly and quietly and sit for lengths of time. Shy animals will reveal themselves only if they think they are alone. Encourage children to use their senses (except taste). Sometimes, listening is as important as looking.

A study of a nature square will provide many opportunities for the use of books and other media as they are available. Handbooks and field guides for identification of insects, spiders, birds, snakes, flowers, and plants will be especially helpful to you.

If they are available, films, filmstrips, or video-cassettes can be of great help in supplementing this area of study.

As you identify insects and spiders of different kinds, classify them into harmful, coexisting, and helpful categories as they relate to man. Although many people dislike them, spiders are usually harmless and are very valuable in the control of harmful insects.

Figures 52-1 through 52-3 show an insect, a spider, and a mouse. These should be helpful for quick classification so children can pursue further identification in field books.

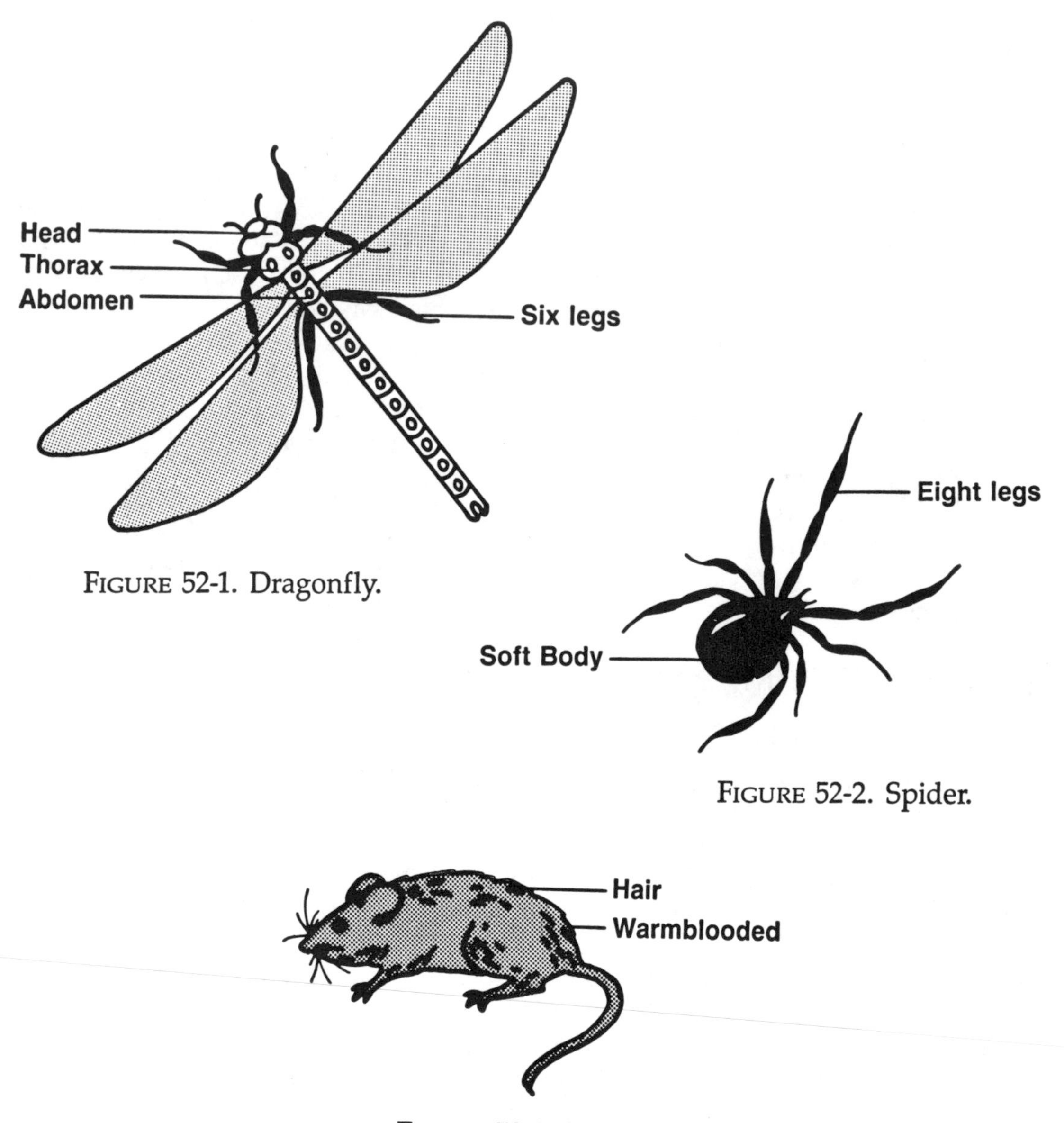

FIGURE 52-1. Dragonfly.

FIGURE 52-2. Spider.

FIGURE 52-3. Mouse.

ACTIVITY 53: How Can We Attract Wild Birds?

MATERIALS NEEDED

- Wooden board 45 cm. long × 15 cm. wide × 3 cm. thick (16″ × 6″ × 1″)
- Two wood slats 45 cm. (16″) long
- Five wood slats 15 cm. (6″) long
- 15 to 20 nails
- Hammer
- Bird seed, suet, bread crumbs, breakfast food, grains, etc.

PROCEDURE

1. The following instructions will help you make a feeder to attract different kinds of birds. Use the two long slats and two of the shorter slats to enclose the long wide board with a rim. Carefully nail the last in place.
2. Arrange the remaining three slats inside the enclosed space about 10 cm. (4 in.) apart so they form partitions to divide the large board into four sections. Use two nails on each end to hold the slats in place. If you live in a rainy climate you may want to construct a roof of heavy plastic and coat-hanger wire. When you finish, your project should look something like Figure 53-1.
3. Find an isolated place near a window of your school building or a nearby home. Try to find a place where there are trees and shade nearby. The feeder *must* be near a tree, large shrub, and some safe place (ledge or rooftop) so birds will have some place to go if they are startled.
4. Locate your feeder off the ground (hanging from a tree limb, pole, or wooden base), low enough so it can be refilled. Be sure to put it in a place where you can observe it from a window.
5 Wait a day or two for your scent to leave the feeder. A shallow bowl with water might be included. Because you will leave a human scent, try not to handle the food more than is necessary.
6. Observe your feeder as much as possible for two weeks. When you see a bird, make a quick sketch of it noting color and size. The next activities will help you identify some of the birds your feeder attracts.

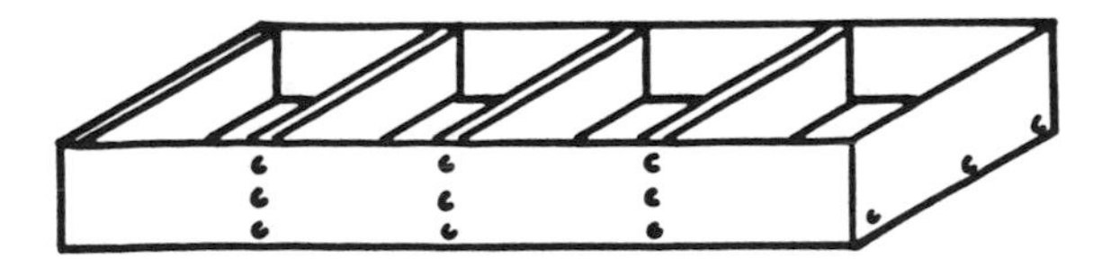

FIGURE 53-1. Birdfeeder with four sections.

TEACHER INFORMATION

Birds are everywhere. Even large cities and arid desert regions have many species. Birdwatching is inexpensive, entertaining, and rewarding. The next activities should help children develop an understanding and appreciation of the many contributions birds can make to our lives.

Note: Do not use glue on birdfeeders or birdhouses. Birds don't like the odor.

ACTIVITY 54: What Is a Bird?

MATERIALS NEEDED

- One drawing of a typical bird per student
- Pictures of many varieties of birds
- Caged live pet bird

PROCEDURE

1. Figure 54-1 shows a typical bird.
2. All birds have these same basic parts but some appear to be different to help the bird to survive in its particular environment. Refer to the typical bird parts as you read steps 3 through 6.
3. Notice the short, pointed beak. This could tell us the bird eats certain things for which a short beak is helpful. If the beak is slightly longer it might belong to a woodpecker. Fishing and shore birds usually have very long beaks. Ducks have wide, flat beaks for sifting water on and below the surface in search of vegetation. Birds of prey often have very sharp, curved (hooked) beaks. A beak is often called a bill. The pelican is famous for its long bill and pouch.
4. The feet and legs of birds often tell much about them. Short legs and feet with short claws usually belong to a land-perching bird (such as the one in the picture). Fatter feet with longer toes often suggest a ground-nesting bird. Most swimming waterbirds have webbed feet while birds of

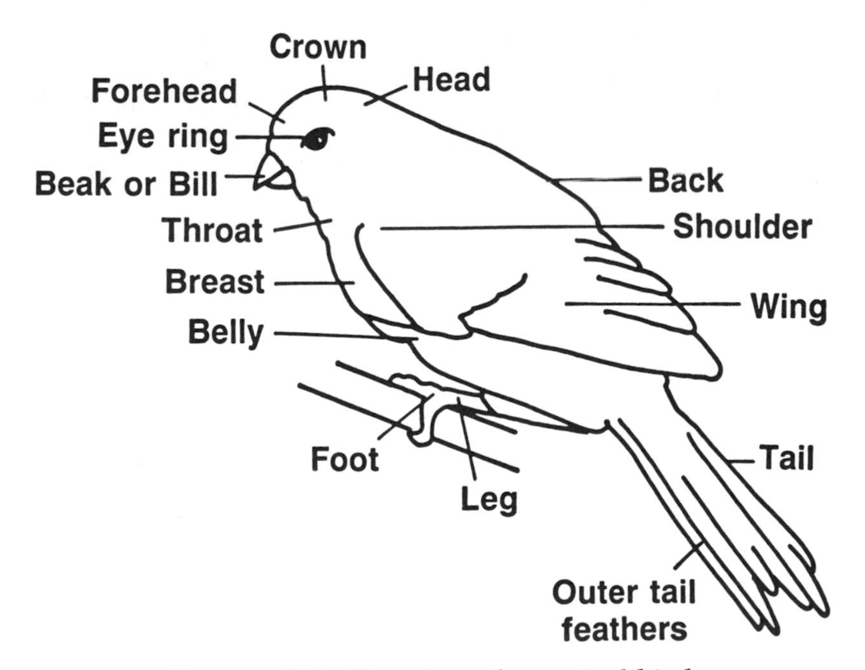

FIGURE 54-1. Drawing of a typical bird.

prey have powerful curved feet with long, sharp claws. Shore and marsh birds usually have longer, thin feet and very long legs. Many birds have a characteristic of standing or resting on one leg.

5. The wings of birds vary greatly, again, depending on their environment and feeding habits. Our typical bird has a wingspan (open wings) more than twice the length of its body. Some soaring birds have very wide wingspans while birds of prey have flatter, wider wings that permit them to both soar and hover.
6. Although our typical bird is not shown in color, colors are very important for bird study. Males usually have brighter colors than females. Female birds are not colored as brightly so they can remain hidden when they are nesting. Different species of birds have distinct markings. Ornithologists (scientists who study birds) use colors and markings as a major means of identifying birds.
7. Choose a bird picture from your teacher's collection. Put it on your desk beside the drawing of the typical bird. Can you find the same parts? From what you have learned about birds, tell all you can about the bird you have chosen.
8. Study the live bird in the cage. In what ways is it different from the ones in the pictures? Tell all you can about the live bird.
9. You are now ready to begin your observation of wild birds.

TEACHER INFORMATION

A high-quality field book for bird identification is essential. Binoculars are helpful especially if your bird feeder is located some distance from the classroom. A notebook and sketch pad should be available at all times.

These activities are designed to introduce children to bird study. Your library or media center can provide many other materials and suggestions. Depending on where you live (remember students will move several times in their lifetimes), birdwatching can become an exciting and interesting lifelong hobby. If your school is not a member, we strongly suggest that you join the National Audubon Society, Membership Data Center, P.O. Box 2667, Boulder, Colorado 80321. There may be a local chapter in your community.

ACTIVITY 55: What Will Attract Birds to Your Home?

(Construction project under teacher supervision)

MATERIALS NEEDED

- Assorted pieces of wood
- Saw
- Hammer
- Plastic and paper milk cartons
- Assorted nails
- Dowels
- Paper

PROCEDURE

1. Figures 55-1 and 55-2 show pictures of simple birdhouses.
2. If you would like to construct one, make a drawing (plan) and show it to your teacher.

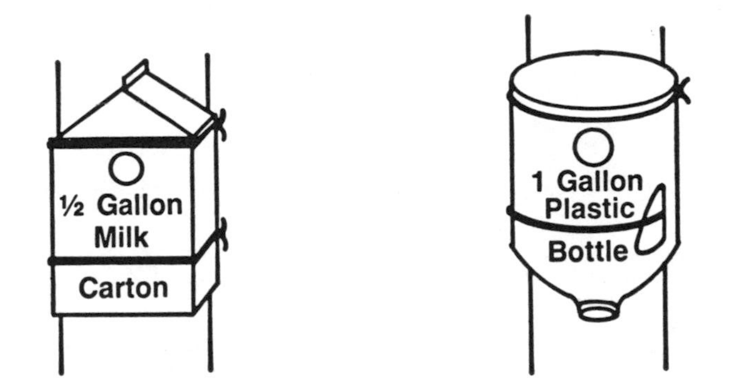

FIGURE 55-1. Two birdhouse designs.

FIGURE 55-2. Birdhouse design with cat guard.

3. Your teacher will help you gather the materials and build a birdhouse.
4. If you prefer, take your teacher-approved plan home and have someone help you.
5. Share your birdhouse with the class.

TEACHER INFORMATION

If there is enough interest, this could become a project for everyone in the classroom. If so, you will need parents or aides to assist.

Don't feel limited to the birdhouses in the pictures. Urge students to be creative in their designs. Remember birdhouses need to be durable and weatherproof. Also, different kinds of birdhouses will attract different birds. The size of the hole, for instance, will help determine the type of bird attracted to the birdhouse. Consult a bird book for specifics. Birds may not move into the houses for some time due to season, scent, location, or other factors.

Commercial birdfeeders and birdhouses are available but should be purchased only if students are unable to construct them. Kits from which birdhouses can be constructed are often available.

If birds are plentiful in your area, birdhouses make wonderful gifts.

ACTIVITY 56: What Do We Need to Keep Pets in an Aquarium?

(Total-group project)

MATERIALS NEEDED

- 10-gallon (or larger) aquarium, glass cover, and stand
- Heater (see step 2, "Procedure")
- Thermometer (see step 2, "Procedure")
- Air pump
- Clean gravel
- Sand
- Water plants
- Water (see step 2, "Procedure")
- Snails
- Fish (see step 2, "Procedure")
- Fish food
- Rocks of various sizes and shapes
- Small fish net

PROCEDURE

1. An aquarium is a place in your classroom for aquatic animals. With your teacher's help you can make the aquarium into a comfortable home. Place the aquarium in a sunny, protected place in the room.
2. Decide on the variety of fish you will want to live in your aquarium. Freshwater fish (goldfish varieties) can usually live comfortably without a special heater in most classrooms. Tropical and exotic fish need a carefully controlled environment, so you will need a heater and thermometer. If you choose saltwater fish and don't live near an ocean, your pet store may have salt you can add to fresh water but it is difficult to maintain.
3. Put 3 cm. (1 in.) of clean gravel in the bottom of the aquarium, sloping upward from front to back. Add assorted rocks near the back.
4. Spread 3 cm. (1 in.) of clean sand over the gravel slope.

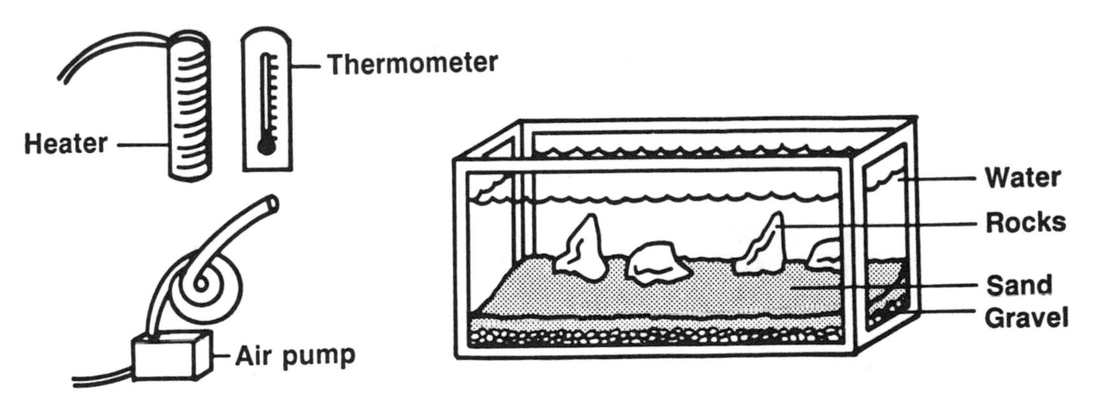

FIGURE 56-1. Aquarium with heater and air pump.

5. Slowly add water (so as not to stir the sand) up to 5 cm. (2 in.) of the top. For freshwater fish, pond water or well water is best. If you use city water, let it stand in open containers for several days before adding it to the aquarium (to remove chemicals).
6. Add green water plants (a variety of six or more). These can be purchased at a pet store and will depend on the kind of fish you choose to keep.
7. If you have chosen to make a freshwater aquarium, all you need to do is add several fish and freshwater snails to keep the tank clean. Put the lid on top. If you have chosen tropical fish, you will need to continue with the following steps.
8. Install the heater and thermometer and wait for the water to reach the desired temperature. (NOTE: Some heaters come with thermostats, but a thermometer is still helpful as a safety check.)
9. Secure the hose from the air pump to a rock on the bottom of the aquarium. Turn it on.
10. Put the tropical fish in your aquarium. Instead of snails you may need to purchase special fish with suction mouths to keep the tank clean. The pet store can provide special instructions for testing the water and feeding fish.
11. You now have a type of "aquatic nature square" in your classroom.
12. Look at the Observation Chart you used to study your outdoor nature square. Using it as a model, design an Observation Chart for your aquarium.

TEACHER INFORMATION

If you have never constructed an aquarium before, talk to a pet store owner or someone else about the special needs of tropical fish. You may need additional equipment to maintain it properly.

Aquariums with tropical fish are very attractive and interesting but freshwater or saltwater aquariums (near the ocean) are more representative of water life as it exists. There are several additional advantages to a freshwater aquarium. If it is 10-gallon or larger, it will be nearly balanced and not require an air pump. "Balanced" means that the plants produce about the same amount of oxygen as the animals use. The animals in turn release enough carbon dioxide for the plants. If you match your fish, snails, and plants properly (this usually takes time), the fish will get food from the plants. The plants will use the waste products of the fish for food to produce new growth, and the snails will keep the whole area tidy and clean.

Occasionally you should remove the glass lid to permit air to get in (or leave a ventilating space), and you may want to add a little fish food to give your pets a treat. An aquarium that is nearly in balance will need a change of water about twice a year (add pond water occasionally to replace evaporated water). When the water is changed, you will need to start at the beginning with fresh, clean gravel, sand, and rocks. Again use fresh pond water. The same plant and animal life can be retained.

Periodically have the students assist you in checking the "health" of your aquarium. Usually conditions will let you know if you need more plants, animals, or snails.

Don't be concerned if a green slimy film develops on the top of the water. This is a sign of a healthy aquarium. Probably plants and animals too small to see with the naked eye are also living in the tank. In a following activity you will have an opportunity to study these microorganisms.

Freshwater aquariums can also provide a home for tadpoles. When they begin to grow legs, simply transfer them to a combined environment of pond water and land and watch them develop into frogs. Use shallow water and tilt the tank so part of the tank floor is not covered with water. This will assure that the tadpoles can get air when their lungs begin to develop, in case they are not transferred quite soon enough. More detailed instructions for raising tadpoles may be found in Section 4, "Growing and Changing: Animal Life Cycles."

ACTIVITY 57: How Can We Keep Small Animals in Our Classroom?

(Total-group project)

MATERIALS NEEDED

- Aquarium or terrarium (at least 20 gallons)
- Clean gravel
- Clean sand
- Potting soil
- Plants (see "Procedure")
- Assorted rocks, small pieces of wood
- Small, shallow plastic pan
- Heavy wire screen
- Reptiles or small animals

PROCEDURE

1. You can keep small animals in your classroom if they have a comfortable place to live. Find a warm, sunny, protected place in your room as a permanent location for your terrarium. Although you may be using the same container, aquarium means *water home,* and *terrarium* means *land home.* Be sure it is clean.
2. With your teacher and other class members, choose the kind of animals (reptiles, amphibians, or mammals) you would like to have. If you have enough containers you could make homes for each variety.
3. Use library books to learn about the animals you have chosen and plan a comfortable environment for them.
4. Put about 3 cm. (1 in) of gravel in the bottom of your terrarium. Add 3 cm. (1 in.) of sand on top. Arrange the rocks and pieces of wood throughout the container to provide privacy and shade.
5. Locate the plastic pan near one side. If you have chosen desert animals they will need very little water. If you have chosen turtles or amphibians you should use a water container somewhat deeper that will cover about one-third of the bottom. Put several flat rocks in the amphibian or turtle container so the animals will have a place to climb out and "sun".
6. Be sure the pan is easy to remove as you will need to clean it regularly. Depending on the variety of animal you have chosen, you may need to add a small tray for food.
7. From the study of the animal you have chosen, select small plants that naturally occur in its environment. If you are housing animals that like a moist environment, mix 2 cm. (½ in.) of potting soil in the sand.
8. Before you put your animals in their new home, be sure you have a heavy wire covering over the top so your animals cannot climb or jump out.
9. Your terrarium cannot be "balanced" as a fresh water aquarium can be; therefore, your animals will need fresh food and water regularly. The plants will need varying amounts of moisture, too.

10. Put your animals in their new home and watch their behavior. Can you design an Observation Chart to help study them?

TEACHER INFORMATION

Resource people can be of great help in this study. If aquariums are not available, cardboard or wooden boxes with the top and one side covered with heavy plastic can be used for insects, reptiles, and amphibians. Be sure to provide plenty of air holes covered with screen. Mammals, especially rodents, will gnaw through cardboard or wood and must be housed in glass or metal cages. Use a sieve to clean the sand in the bottom of the cage about once a week. Commercially purchased cages are excellent for temporary housing. They are easy to clean and

FIGURE 57-1. Desert terrarium.

FIGURE 57-2. Woodland terrarium.

FIGURE 57-3. Pond terrarium.

reduce the amount of care required; however, it is difficult to create a realistic environment in them.

Housing male and female mammals together can become a problem. If the female becomes pregnant she will need to be separated from the male. Pregnant rodents are not recommended for classroom care and study. During gestation they should not be disturbed for cleaning (especially their nests) and after birth, if frightened or molested, they may eat their young.

If you construct an amphibian environment you will be able to start with tadpoles and watch them change into frogs. See specific instructions in Section 4, "Growing and Changing: Animal Life Cycles."

Turtles, newts, and frogs make very good amphibian-reptile pets. Horned lizards, common lizards, tortoises, and small snakes will live together in a desert environment.

Usually it is best to have only one species of mammal. Gerbils, hamsters, and guniea pigs are clean, and easy to care for. Wild mice, rats, or minks are not recommended.

Many insects will live together. Spiders (including the gentle tarantula), beetles, and crickets can share the same cage. Praying mantises are very interesting but should be housed alone, as they are voracious eaters and also cannibalistic. Books and resource persons can help in your selection of animals.

When your terrariums are complete, have the children list the "daily care" tasks and organize groups with specific time, day, and date assignments.

Live specimens and food can be obtained by collecting, and from pet stores and science supply houses.

If animals are kept in the classroom, some are likely to die there. Teachers need to be aware of the sensitivity of this issue with children. Insects have a short life span. Insects and spiders eat other insects and spiders. Tadpoles have a high mortality rate from egg to frog. Handled properly, these can be valuable learning experiences for children as they observe life's natural processes.

ACTIVITY 58: What Very Small Animals and Plants Live Around Us?

MATERIALS NEEDED

- Biologist or naturalist who studies microorganisms in water (not bacteria)
- Equipment furnished by specialist
- 9″ × 12″ drawing paper
- Pencil

PROCEDURE

1. There are many tiny animals, too small to see with the naked eye, that live around us. You have already used magnifying lenses to examine some small insects and plants. In your study of birds you may have used binoculars. Today we are going to see tiny animals that require very powerful lenses to be seen.Scientists call these lenses microscopes and microprojectors. Figure 58-1 shows some of the most common organisms you may see.
2. Across the top of your paper, draw pictures of the animals in the figure. As you see other animals in the microscope today, try to make a drawing of them. Put your pictures under the drawings of the ones in the figure they most closely resemble. You are now ready to conduct your inquiry.
3. After you have seen and made drawings of the tiny animals and plants, share your findings with the class. How are they alike? How are they different? Which did you like best?

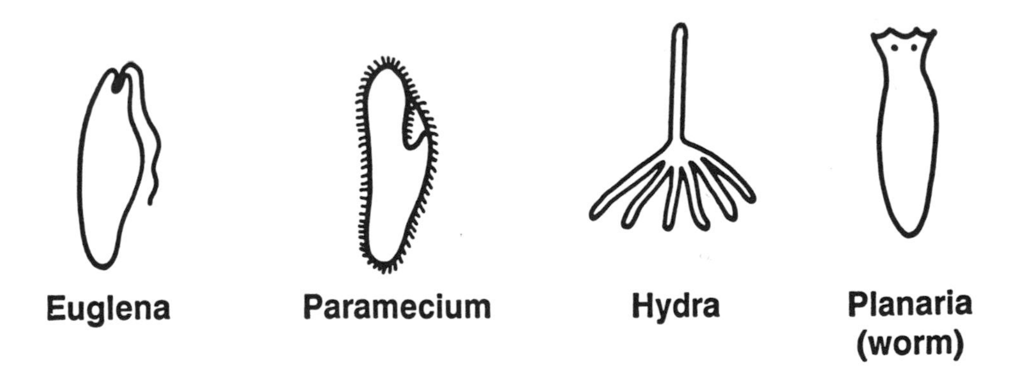

FIGURE 58-1. Drawings of four microorganisms.

TEACHER INFORMATION

If you have access to a low-power (good quality) microscope or microprojector and books on microorganisms, undertake this study alone. It is not necessary to make positive identifications of every organism you find. Use the figure to make general classifications.

Microorganisms are usually present in pond water, the surface of your freshwater aquarium or in a *hay infusion,* which you can make or purchase from a science supply house. A simple film on organisms of a pond may be helpful.

If you invite a specialist, be sure to explain that you are interested in a simple, general introduction with many visuals. The purpose of this activity is to develop awareness of these tiny creatures, not to learn all about them. The students will encounter courses in later years where they will be able to explore in depth.

An in-depth study of microorganisms could be an enrichment activity for highly motivated students.

ACTIVITY 59: How Do Earthworms Live?

MATERIALS NEEDED

- One quart or larger wide-mouthed clear jar
- Tall, thin tin can with one end closed
- Rich, moist soil and sand
- One sheet heavy black paper
- Earthworms

PROCEDURE

1. Study your earthworms. Can you find the head end? Can you find the tail end? How do they move? What surfaces do they seem to like best? Do they seem to like water? Do they like light? What else can you discover about your worms?
2. Put the tin can, closed end up, in the large jar. Fill the large jar with soil, level with the top of the tin can. Add a thin layer of sand.
3. Put your earthworms in the jar.
4. Cover the outside of the jar with black paper.
5. Leave the jar covered for at least 24 hours.
6. Uncover the jar and examine the results.
7. Can you see evidence that the earthworms have been at work?
8. Study the sand on top of the soil. Earthworms take soil through their bodies, remove the food, and deposit a substance called castings.
9. Cover your jar for another 24 hours. Repeat your study of your worms.
10. Write or tell everything you can about earthworms.

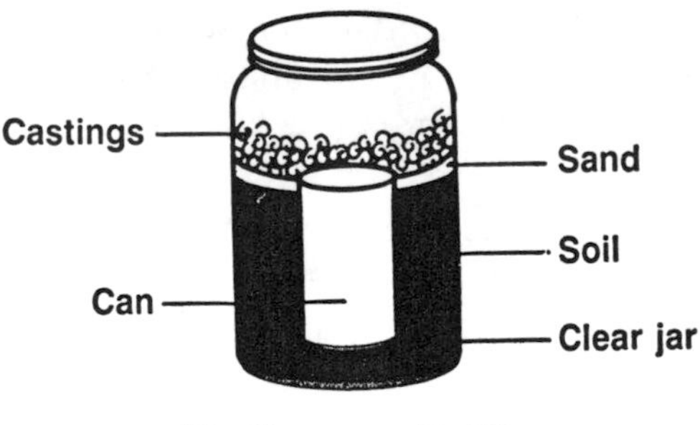

FIGURE 59-1. Jar and tin can.

TEACHER INFORMATION

Earthworms will not tunnel near the surface of the jar until it is covered with heavy black paper to make it dark. The can is used to make the layer of soil inside the jar thinner and force the earthworms to tunnel near the surface of the glass jar. If you decide to continue the activity for several days, the soil should be kept moist. Take care not to overwater.

ACTIVITY 60: Why Is the Snail So Unusual?

MATERIALS NEEDED

- Snails (one per student)
- Black construction paper (one per snail)
- Flashlight
- Newsprint
- Magnifying glasses
- Small pieces of lettuce

PROCEDURE

1. *(Read this before you begin):* The snail is a most magnificent animal. It is shy, yet curious, calm, quiet, usually helpful, and it doesn't disturb others. It is creative and artistic, slow but hard working, and tidy; it doesn't depend on others for its home or food (some even carry their homes around with them on their backs). Snails glide through life as vagabonds. Sometimes they become too plentiful and damage or eat people's food, but when that happens, nature reduces their numbers and soon they roam free once again.
2. Choose a snail and put it on a piece of black paper. Give it a name so you can talk to it. Study your snail and compare it with Figure 60-1. Can you find the same parts?
3. Because snails are shy, you may have to wait a little while for it to come out of its shell and begin moving about.
4. Notice its head with the round-tipped feelers or tentacles sticking out. Land snails have two long tentacles with eyes at the end and two short, stubby ones that some scientists believe help them smell. Carefully touch one of the feelers and watch what happens.
5. Examine the underside of your snail. The whole bottom is a large foot. The front of the foot secretes a slippery substance called mucus on which it can slide. Since the mucus is also sticky, the snail can go straight up or down without a problem.
6. When your snail begins to come out of its shell, put a small piece of lettuce about 15 cm. (6 in.) away. Be sure it is on the black paper.

FIGURE 60-1. Snail with feelers, head, and foot extended.

7 Watch your snail to see if it can see or smell the lettuce. If it begins to move toward it, measure how far it can travel in one minute. Look at its "glide" path on the paper.

8 Make a wet spot on the paper. Is the snail attracted to it? Snails must have moisture in order to survive. They prefer a shady, moist environment. On hot, dry days they go into their shells to wait for moisture.

9 Darken the room and shine a flashlight near your snail. Does it move toward or away from the light?

10. Think of other investigations to conduct. Don't do anything that might harm your snail.

11 Share your findings with the class. Who has the fastest snail? How could you find out?

TEACHER INFORMATION

Snails are plentiful in many parts of the country. Many of the same activities can be adapted for earthworms or meal worms. The snail completely fills its shell and as it grows larger it adds additional swirls. Slugs are "naked" snails, with the same characteristics, but they usually need more moisture.

An average snail, under "full steam," travels about 5 cm. (2 in.) in a minute.

Although they are sometimes considered pests, snails eat decaying vegetation and can improve the soil in a garden.

Water snails eat decaying vegetation to assist in the balance of a pond or ecosystem. They will keep your aquarium clean in the same fashion.

Section 4

GROWING AND CHANGING: ANIMAL LIFE CYCLES

TO THE TEACHER

This section, "Growing and Changing," should follow or be integrated into the preceding section, "Animals." Depending on where you live, the "Animal" section may be taught in early fall, but in locations where distinct seasons occur, "Growing and Changing" is best taught in the spring months.

Several of the activities involve the use of live specimens. If you live in an area where you cannot collect caterpillars or tadpoles, you will need to place orders with pet stores or biological science supply houses in January or February. If you order early and specify a later shipping date, most supply houses are happy to cooperate. Be aware that some states require USDA permits for interstate shipment of live animals. The supply catalog will advise you if this is necessary. Animals purchased in local pet stores will already have been approved.

Obtaining, organizing, and filing a wide variety of pictures is essential to quality teaching in the elementary school. In addition to the journals recommended in the appendix of this book, old biological science supply catalogs obtained from a local high school or college should be of great help. Many are filled with high-quality photographs. Obsolete science textbooks and library books can often be obtained from your school district free of charge. The text material may be obsolete but the pictures often are not. Some organizations, such as the Audubon Society, sponsor junior organizations for young people.

In some activities in this section, animals that lay eggs with the young developing and hatching outside the female's body are studied. Others utilize animals that retain the fertilized egg inside the mother's body. The terms *hatching* and *giving birth to living young* are used in some manner to make this distinction.

Instructions are included for hatching eggs in the classroom. Attempting to show living birth at the elementary level is *not* recommended. Smaller mammals, such as mice or gerbils, should not be disturbed during the gestation time even for cage cleaning. After birth, if mothers are frightened, they might eat their young.

Integrated creative learning activities are suggested throughout this section. Try to include as many as possible.

ACTIVITY 61: What Is Growing Up?

(Teacher-directed total-group activity)

MATERIALS NEEDED

- Chart or blackboard with the following terms listed: newborn infant, baby, "toddler," young child, youth, adolescent, "teen-ager," young adult, "grown-up," adult, mature adult, "senior citizen" and old person
- Lined paper
- Pencils

PROCEDURE

1. On the chart or blackboard are some words we often use to describe humans of different ages. Discuss each word with your teacher and the class. Do some words refer to the same age group? If so, choose one word for each group.
2. Write the words down the left side of your page.
3. Put numerals by the words to show how old you think people must be to be classified (put) in a certain group.
4. Next to the words write the name of someone you know in each age group. Share your list with others in the class.

TEACHER INFORMATION

During the group discussions you will need to add and change terms until students are able to associate someone they know with each age group. If you have not done this before, be prepared for some surprises. Research indicates that young children's concept of age is similar to that of time and seems to develop as they mature. Because they are "today" oriented, infancy seems long ago and "when I grow up" can mean almost anything. As they grow older, students develop a greater awareness of their progress into the next age group and some feelings for the expectations of adult life.

Activities 61, 62, and 63 attempt to establish some understanding of pre-birth or gestation, babies, children, and adults. In addition to humans, some other animals in various stages of development are introduced in Activities 62 and 63.

You may discover that young children will recognize only four or five categories: babies, children, adolescents, adults, and old people. They may decide adults and old people are the same. The concept of growth and change can be developed with four basic categories. Aging is common in natural animal populations. Old age is rare.

ACTIVITY 62: How Does Growing Up Begin?

MATERIALS NEEDED

- Pictures of humans, both male and female, of all age groups especially mothers with babies
- Pictures of pregnant women
- Pictures of mammals of all kinds (some pregnant females if possible)
- Pictures of mammal babies being cared for by their mothers
- List of age groups developed in Activity 61

PROCEDURE

1. Study the pictures of humans. Group them according to the age classifications you chose in Activity 61. You should have at least one picture for each group.
2. Can you find women in the pictures who will have babies soon? How can you tell?
3. Look at the pictures of animals that are not humans. Put them in the age classifications with the people. Do you have an animal picture in each group?
4. The animals you have classified are called mammals. Like humans, the females all carry their babies inside them until they are ready and then give birth to living young. Can you find pictures of animals you think will soon give birth to babies? How can you tell?
5. Compare the pictures of human babies and other animal babies. Notice that they need their mother's care.
6. With your teacher and other members of the group, discuss how humans and other mammals care for their young.

TEACHER INFORMATION

This activity begins with an overview of humans and alludes to a relationship with other mammals. The focus then narrows to their common method of giving birth, and their extended care of the young. It is not intended to introduce a study of reproduction except as it applies to the beginning growth process. Reproduction is essential to the perpetuation of the species. All living things have the ability to reproduce themselves.

Mammal babies are born in a relatively helpless condition and require care from their mothers or some other mature adult.

Concepts to be emphasized in this activity are:

1. Human and other mammal young begin as eggs and grow and develop inside their mothers. They are quite small at birth.
2. When compared to most other animals, human and other mammal

babies are quite helpless at birth and require care for longer periods of time.

Tell your students that the role of the male in caring for the young has been intentionally excluded, as it varies so greatly from species to species.

An interesting extension of this activity is the study of marsupials. Live young are born in an early stage of development and spend time in a special pouch in their mother's abdomen, consuming milk until they are fully developed. Although students may be more familiar with kangaroos, the opossum is the only native marsupial found in the United States.

ACTIVITY 63: Why Do Some Animals Hatch from Eggs?

(Teacher-supervised small groups)

MATERIALS NEEDED

- Hard-boiled chicken eggs (one per group)
- Squares of paper towel (one per group)
- Dull metal or plastic knife
- Drawing of cross section of a chicken egg (one per student)

PROCEDURE

1. Eggs that hatch outside the mother's body are a part of the reproductive process of many animals. Study your diagram of an egg. Use it to locate the parts in a real chicken egg.
2. The chicken egg has been hard-boiled so the contents will not run out. Place the egg on the square of paper towel and gently crack it in many places by tapping in on a table or desk.
3. Carefully peel the hard outer shell from the egg. Be sure to keep the shell on the paper towel.
4. Look at the egg without a shell. Can you find the thin membrane?
5. Use the diagram to help find the place where there was an air space between the shell and the soft white part of the egg. How did you find it?
6. Use the knife to carefully cut the egg in half.
7. Compare your diagram with parts of a chicken egg. Did you find them all?
8. Discuss your findings with your teacher and other class members.

TEACHER INFORMATION

The egg diagram has been simplified to show basic parts that students can easily identify. For this activity, unfertilized eggs are preferred.

Many of us are familiar with an egg as food but have never carefully examined its parts. The hard outer shell serves as a protective covering. Some animal eggs do not have protective outer coverings, or if they are present they may appear in a different form, such as the soft leather-like shells of reptiles.

A thin membrane encloses the albumen and yolk. If the egg is fertilized, material in the albumen will form the embryo. The yellow yolk provides food for the embryo until some time (at least 24 hours) after it hatches. Birds hatching from hard-shelled eggs have a special "egg tooth" on their beaks to assist them.

Although they appear to be solid, egg shells are porous to allow the exchange of oxygen and carbon dioxide necessary for a developing embryo. Chalaza are whitish spiral bands extending from the yolk to the membrane at each end of the egg. You may not be able to identify them in a hard-boiled egg. You may substitute a raw egg in a small bowl. The parts shown in Figure 63-1 may be easier to observe. Be careful of spills.

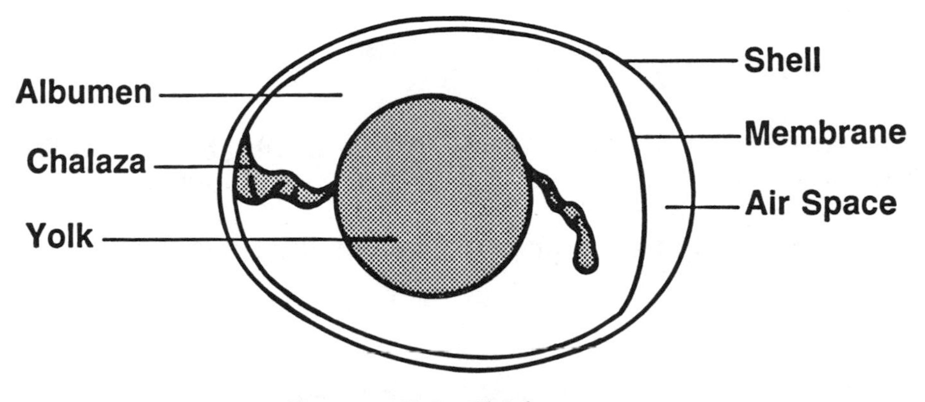

FIGURE 63-1. Chicken egg.

ACTIVITY 64: What Are Some Other Animals That Hatch from Eggs?

(Groups of four or five)

MATERIALS NEEDED

- Pictures of many different animals hatching from eggs: fish, amphibians, tadpoles, reptiles (snake, crocodile), insects (grasshoppers, caterpillar), spiders, and birds
- Pictures of adult metamorphosed animals: butterfly or moth, and frog

PROCEDURE

1. Look at the pictures of different kinds of animals hatching from eggs. Many look somewhat like their adult parents. Some do not.
2. Classify the pictures into two groups: (1) those that look like their parents and (2) those that do not.
3. Ask your teacher to check your groupings.
4. When certain animals hatch, they go through several changes before they become adults. Changing through several stages of growth is called *metamorphosis*. Group two should include pictures of these animals in various stages of development.

TEACHER INFORMATION

The major purpose of this activity is to introduce the concept of growth and change through metamorphosis. Activities 66 and 67 will develop the concept in greater depth.

It is also important to observe that many insects, fish, reptiles, and other animals that resemble their parents when they hatch also possess instinctive behavior that resembles that of the adult. Reptiles and many fish are examples.

Many adult animals lay their fertilized eggs in a place where food will be available to the hatching young and then abandon them. The term "baby" is most often applied to humans and other mammals. To many children, "baby" also implies the presence of a mother. Since mothering is not instinctive among some animals, scientists use many different terms to identify and classify immature and mature animals.

ACTIVITY 65: How Can You Hatch Eggs in the Classroom?

MATERIALS NEEDED

- Fertilized chicken eggs
- Garden spider eggs
- Praying mantis eggs
- Incubators (see "Teacher Information")

TEACHER INFORMATION

Hatching eggs in the classroom can be an interesting and exciting first-hand experience for students. However, there are problems involved. This activity has been listed as optional, based on your interest and materials available. Following are two examples of egg-hatching activities. Please read carefully before you begin.

1. *Hatching chicken eggs:* A major problem with hatching bird eggs is the 21-day incubation period. Several biological supply houses offer incubators but costs are fairly high. Several science textbooks listed in the bibliography provide plans for building simple incubators using boxes, wire, and incandescent lights. Plan to use an incubator large enough to accommodate several (up to 12) *fertilized* eggs, as usually some will not hatch (tiny commercial incubators that hold one or two eggs are not recommended).

You will need a thermostat to ensure a constant temperature in the incubator throughout the 21-day period. Be sure the electric power in your school is not turned off overnight or on weekends.

Eggs must be turned several times (three or four) every 24 hours. Some commercial incubators do this automatically. If you are planning to do it manually, be sure to mark the eggshell so students will know how far to turn it. This must be done on weekends as well as on school days.

The eggs will hatch when the chick is ready, usually *not* during science time. Your eggs may not take a full 21 days to hatch, as they were probably fertilized and laid a day or two before you obtained them. Incubating a large number of eggs increases the possibility of having some hatch during the school day.

Fertilized eggs may be obtained from hatcheries or biological supply houses. As an alternative to chickens, quail or pheasant eggs may also be available.

If incubation is not practical, some hatcheries may be able to provide eggs that are ready to hatch within a day or two. You will still need an incubator but the time involved is much less.

Be sure to prepare the students for the experience in advance. Your library or media center can provide books and pictures.

If you live in or near a rural area you may be able to borrow a mother hen with fertilized eggs. She is a natural incubator. She will turn the eggs, keep them warm, and provide care for the chicks after they hatch. If you get a "setting" hen, be sure to have an adequate cage and preparations for feeding and cleaning.

Young chicks hatched in an incubator can remain in it for several days. They will require warmth, water, and food. As they become larger they will need to be transferred to a larger warm space called a brooder. If possible, this is the time to return them to the hatchery or farm. It is not recommended that baby chicks be sent home with children, especially in urban or suburban areas.

Cautions:

a. Not all eggs will hatch and some chicks may not survive the hatching process or may die soon after. Students should be prepared for this eventuality. Under no circumstance should you "help" the chick during the hatching process. The struggle it goes through is part of its beginning life process.

b. A few older science books suggest incubating a larger number of eggs and opening one every other day to observe the development of the chick embryo. This has been found to be inappropriate for elementary school age children and is not recommended.

2. *Watching insect or spider eggs hatch:* Clusters of eggs of the praying mantis may be ordered from biological supply houses. Spider eggs can be collected in the spring. Although neither type of egg requires particular care during incubation, problems may occur after hatching.

Spider eggs should remain attached to the object on which they are found (branch, leaf, or stick). If you bring them into the classroom, be sure to provide a secure cage of tightly woven mesh, as some species of spiders are very small when they hatch. You may need magnifying glasses. It is not necessary to have a female adult spider with the eggs. Release the young spiders in a day or two near the location where the eggs were collected. Most spiders do not or cannot bite humans. A few species can and do (scorpions and black widows). Spider bites are rarely fatal but they can be irritating and painful. For most spiders cotton or leather gloves will provide adequate protection. The book *Charlotte's Web,* available in most libraries, could be used to develop positive attitudes toward spiders.

Praying mantis egg clusters may be ordered from late winter through spring. Plan to put them outside in a shady, protected area and check them daily. As the eggs hatch you will probably be able to catch the small nymphs and cage them in the classroom. They should be housed separately and fed according to the instructions that accompany them. They cannot bite humans, even when they grow to a large size, 5 cm. to 8 cm. (2 in. to 3 in.); however, they are voracious eaters and are cannibalistic.

Local variety stores and biological supply houses often stock craft kits consisting of parts to construct adult animals including the praying mantis.

Spiders and predatory insects such as the praying mantis, ladybird beetles, and dragonflies are valuable in controlling harmful insects. Unless they are dangerous, they should not be killed or disturbed.

ACTIVITY 66: What Is a Tadpole?

(Individual and group activity)

MATERIALS NEEDED

- Pictures from group 2 of Activity 64
- Tadpoles (see "Teacher Information")
- Magnifying glass
- Pencils
- Crayons
- Newsprint

PROCEDURE

1. We have found that some animals do not resemble adults when they begin their growth cycle. Look at the group 2 pictures from Activity 64. Find the picture of a tadpole and a frog.
2. Frogs belong to a family of animals called amphibians. They hatch from eggs and begin the first active phase of their lives in water. There are tadpoles in a bowl in your room. Use a magnifying glass to examine them.
3. Make a drawing of a tadpole on your paper. With your drawing try to show the following:
   a. How does it move about?
   b. Can it see? How do you know?
   c. What body parts can you identify? You may need to use words such as head, tail, legs, fin, body, abdomen, and gills.
   d. Use your crayons to show the color of the tadpoles.
4. Just like other animals, tadpoles need a good environment in order to grow and remain healthy. With your teacher and other class members plan the type of environment your tadpoles will need.

TEACHER INFORMATION

In early spring, frog eggs and tadpoles can be collected from ponds and marshy areas around lakes. If you live in a geographic area where live collection is possible, students may be able to gather their own specimens. They may need the help of parents and other adults. A local high school or college biologist could be of great help in locating and identifying your specimens.

Tadpoles may be ordered from biological supply houses or local pet stores that sell fish. Place your order in early spring. Allow four to six weeks for delivery. Be sure to order a rapidly developing variety such as *Xenopus*, as the tadpoles may require several months to develop into young frogs. Order 30 or more since the mortality rate is high. Plan this activity for early spring.

Expense, time, and other problems in your particular school may make it impossible to carry out this and the following activity as outlined. Kits containing eggs, plastic aquariums, food, and instructions are available from the commercial sources mentioned above.

Before you begin Activity 67 you may want to review Activity 57 in the preceding section, on "Animals." The pond terrarium illustrated in Figure 57-3 is somewhat similar to the environment recommended in the following activity and might be adapted.

ACTIVITY 67: What Is a Good Environment for Tadpoles?

(Teacher-directed individual or group activity)

MATERIALS NEEDED

- Four transparent shoe boxes
- Four blocks of wood 5 cm. (2 in.) tall or four wide strips of corrugated cardboard folded as wedges 5 cm. (2 in.) high
- Pencils
- One gallon (4 liters) of pond water (needed every other week)
- Dipping net
- Thirty or more tadpoles
- One copy of Figure 67-1 per student
- One copy of Figure 67-2 per student

PROCEDURE

1. In the "Animals" section, you learned about aquariums and terrariums for small land animals and fish. Review the illustrations of the aquarium and pond terrarium in Activities 56 and 57.
2. Tadpoles begin their lives resembling fish in many ways and gradually develop into adult land animals. To do this, they must have an environment that combines land and water. Figure 67-1 shows the change from egg to tadpole to frog. Compare your live tadpoles with the ones in the

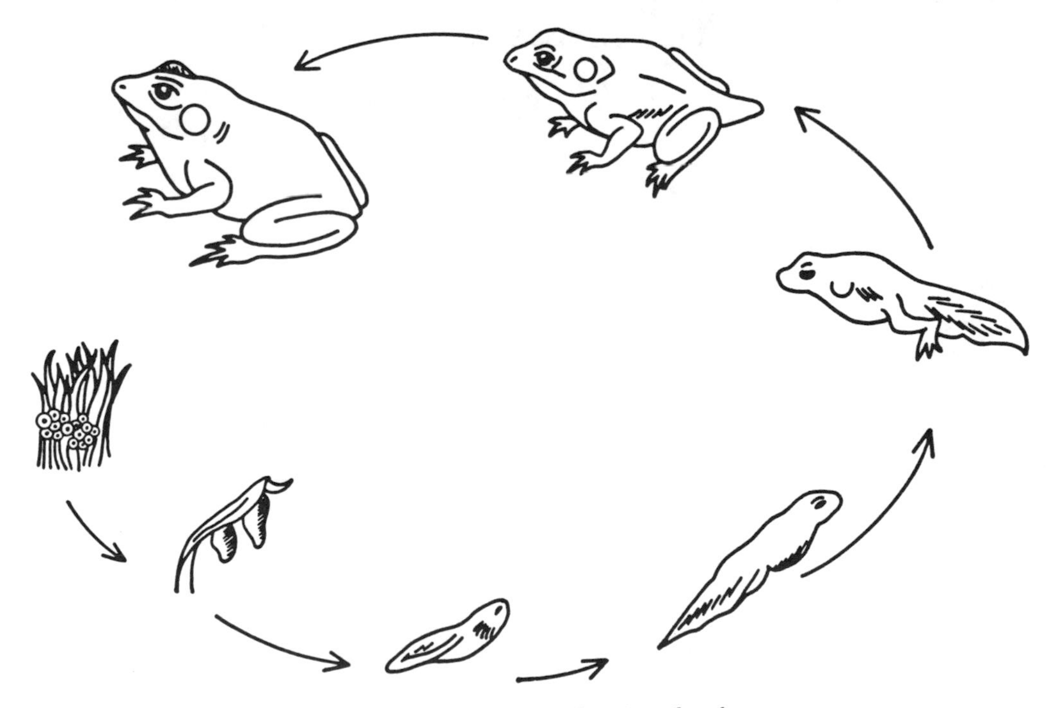

FIGURE 67-1. Metamorphosis of a frog.

drawing. Circle the animal that most closely resembles the live tadpoles. Try to decide whether they are very young or are starting to change. This will help you decide what to feed them.

3. Remove the lid from a plastic shoe box. Most regular shoe boxes are about 35 cm. long × 12 cm. deep (14 in. × 4½ in.). Place the box on a flat surface and use a wedge made of corrugated cardboard or wood to raise one end about 5 cm. (2 in.).
4. Add pond water until more than one-half but not all of the bottom is covered. Your pond water may have a green scum and tiny pieces of material in it. These are healthy signs. Don't remove them.
5. Compare your shoe box aquarium with Figure 67-2. If it is similar, use a dipping net to add several (5 to 8) tadpoles.
6. You will need to change the pond water every other week. If they are very young, tadpoles will eat *small* amounts of boiled spinach. After a few weeks, they eat spinach without boiling. Don't overfeed! Once or twice a week is enough.

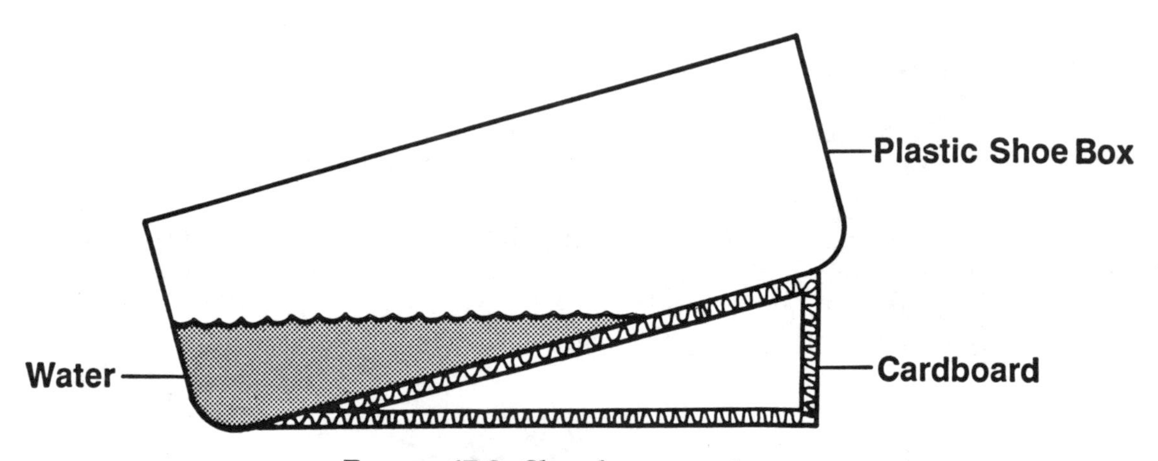

FIGURE 67-2. Shoe box aquarium.

TEACHER INFORMATION

In nature, animals often lay many eggs so that a few will survive to adult stage. With careful feeding and excellent care, you can expect some tadpoles to advance through several stages of metamorphosis.

Tilting the shoe box so shallow water and land are availbale is most important, as the tadpoles lose their gills and become lung breathers. This is usually at about the time the front legs begin to develop. Without shallow water many will drown.

If you teach in an area where pond water is not readily available, try to obtain some from a high school or college biology department or a local pet store—or you may want to create your own as outlined in Activity 56 in the section on "Animals".

Remember, depending on the species, complete metamorphosis in an amphibian may take from four months to over a year. This is a very long time, especially for young children who expect immediate reinforcement.

After the tadpoles are in the shoe boxes put them in a shady place in the room and set up a feeding and water changing schedule. Use the dipping net to remove any dead specimens daily. Discuss the tadpoles briefly as observable changes appear. Remind the children that they, too, take a long time to grow and change.

Be sure to have books and stories about frogs and toads available. Don't put lids on the shoe boxes. Save them to use when you free the animals.

ACTIVITY 68: How Do Butterflies Develop?

MATERIALS NEEDED

- Larva, cocoon, or chrysalis of moth or butterfly
- One copy of Figure 68-1 for each student
- One copy of Figure 67-1 (see Activity 67) for each student
- Insect cage (see Figure 68-2) or butterfly kit from pet store or supply house
- Magnifying glasses
- Pencils

PROCEDURE

1. Study Figure 68-1. It shows the way butterflies and moths develop.
2. Compare this figure with Figure 67-1 in Activity 67. How are they alike? How are they different?
3. There is a new animal in the room in a stage of metamorphosis. Look at Figure 68-1 and find the stage of development your new animal most closely resembles (looks like).
4. If your animal is in the larva or pupa stage, there are usually ways to tell whether the adult will be a moth or a butterfly. Look at Figure 68-1 and see if you can find any differences. Beside the picture of the adult, write the name of the variety of insect you think your animal will become.
5. When your insect becomes an adult, it will soon be able to fly. Do you have a good cage for it? Figure 68-2 shows several simple cages you can make. Choose one and form a group to construct it.
6. Give your adult insect a name. Use books, pictures, and your own observations to learn all you can about it. Notice how many beautiful ways butterflies are used in art.
7. After a few days, take your butterfly or moth to a sunny place near flowers. Say goodbye to it and set it free.

TEACHER INFORMATION

This is a middle-to-late spring activity in many parts of the country. The metamorphosis of a butterfly or moth is an exciting worthwhile experience for students of all ages. The time it takes to develop from larva-pupa to adult is less than that for a frog. Allow 30 to 45 days; however, the larval stage of caterpillars varies before they spin a cocoon or form a chrysalis. Specimens collected locally are best. You will need to provide ample food and moisture for a period. Your encyclopedia or library books will provide detailed instructions.

Occasionally a student may bring a cocoon or chrysalis to class. If this occurs you may want to try to find a caterpillar to use for comparisons with Figure 68-1, butterfly and moth metamorphosis. In the larva stage moth caterpillars are usually hairy or "woolly." Butterfly caterpillars are usually smooth-skinned.

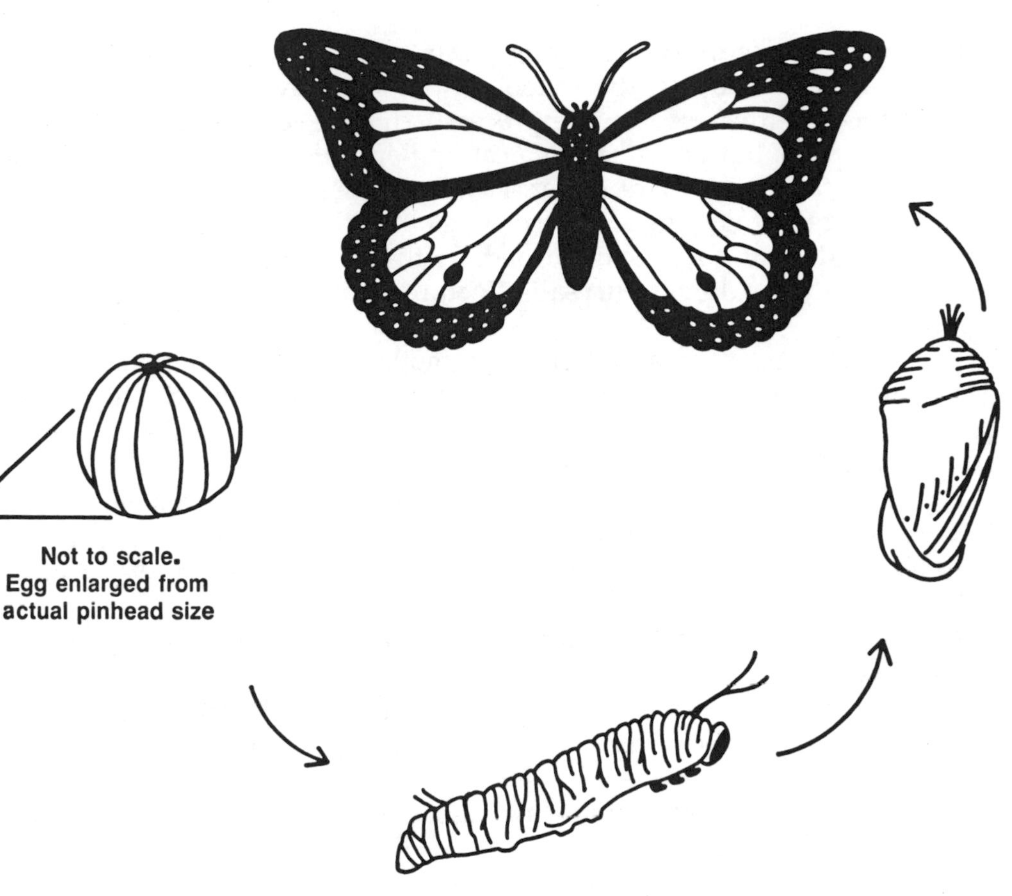

FIGURE 68-1. Butterfly or moth metamorphosis.

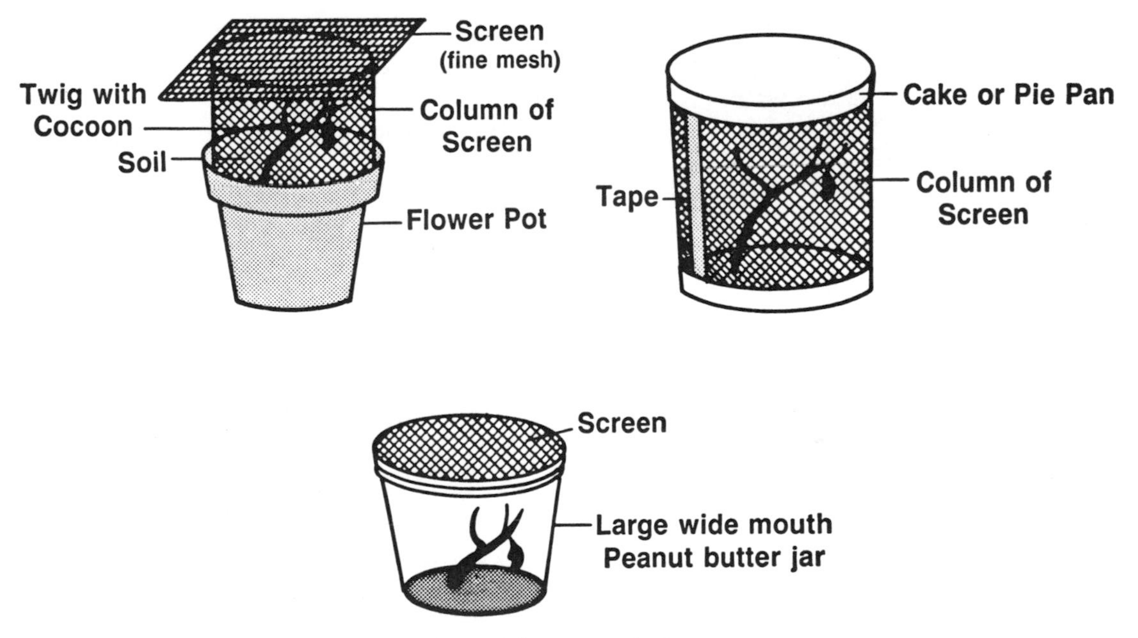

FIGURE 68-2. Homemade insect cages.

Remember, as with egg hatching, your insect may not spin its cocoon or emerge as an adult during the school day If you have several specimens the chance of the children's seeing the process is increased. Mortality rate is lower than with tadpoles but occasionally a caterpillar will die or not emerge from its cocoon or chrysalis. Be patient; never try to help the adult emerge from the pupal state. The struggle to emerge and the drying out of the wings is part of the natural life process and should not be disturbed. Moths spin cocoons. Butterflies form a hard, smooth shell called a chrysalis. Pictures from books may help show the differences.

The development of a butterfly or moth is so exciting, beautiful, and "mysterious" that every child should experience it at some time during the school years. Even if the activity is repeated several times over the years, students will view it with different backgrounds of experience.

If you live in an area where butterflies are not readily available, larva, and even complete kits with cages and food, can be ordered from biological science supply houses at a very reasonable price. However, if possible, we recommend that you collect, house, and care for a local insect.

Tip: Make a butterfly net similar to the one described in Activity 51 in the "Animals" section.

Books, pictures and stories about butterflies are plentiful. Be sure to use them to enrich this experience. Music, art, poetry, and creative movement are important examples.

Collecting, killing, and mounting insects is *not* recommended for elementary students.

ACTIVITY 69: What is Wormy?

(Small-group activity)

MATERIALS NEEDED

- Mealworms in container labeled "Mealworms"
- Earthworms in container labeled "Earthworms"
- Waxed paper
- Figure 68-1 from Activity 68
- Lined paper
- Pencils
- Magnifying glasses
- Tablespoons
- Pie tins or small cake pans

PROCEDURE

1. Have you ever heard someone say "Oh, it has worms in it"? On your lined paper write the names of things they might be talking about. Most of the items on your list are things animals eat. What is similar about them?
2. Line a pie tin or cake pan with waxed paper.
3. Use a spoon to transfer one animal from the container marked "Earthworms" and one from the container marked "Mealworms".
4. Use your magnifying glass to study both worms.Make a simple drawing of each.
5. On your paper write ways they are alike and different. Compare such things as size, color, how they move, legs, head, eyes, mouths, and any other differences you can find.
6. Compare your pictures and description of your worms with Figure 68-1 from Activity 68.
7. One of your worms is not a real worm at all. Can you tell which one? How?
8. Share the findings of your group with your teacher and the class.
9. From what you have learned can you answer this question: Is the silkworm really a worm?

TEACHER INFORMATION

There are many more known species of insects in the world than all other species of animals and plants combined. Many species go through a complete metamorphosis where the larvae bear some resemblance to true worms.

Many insects lay their eggs on or in living or once-living material including plants and wool. When the eggs hatch, the larva feed on the material upon which they were laid. "Wormy" apples are actually apples that have or have had the larva or maggot stage of an insect.

Maggots are legless, softbodied larvae usually found in decaying material. Larvae from the housefly, the mosquito, and relatives are common throughout the world. We often try to control harmful insects at the larva stage (for example, a lid on a garbage can or treating standing water may kill more flies or mosquitoes than you could swat in a lifetime).

A mealworm is the larva of one of several grain-eating beetles. They are available at most pet stores since they are used as food for larger animals (fish, amphibians, and lizards). They are easily stored in a can with bran flakes or a similar food substance. Add moisture to the can occasionally and dispose of the contents at least once during the school year. Adult beetles should be destroyed (privately).

Your earthworm is a true segmented worm and belongs to a completely different group of animals (see your encyclopedia for further information).

Enrichment: Which can move faster—a snail, an earthworm, or a mealworm? How could you find out?

ACTIVITY 70: How Long Do Animals Live?

MATERIALS NEEDED

- Chart of animal life spans (see Figure 70-1)
- Resource people

PROCEDURE

1. Look at Figure 70-1. It shows the average length of life of some common animals. On the left side of the diagram from bottom to top are numbers of years. Across the page are drawings of different animals. Find the animals with the shortest and the longest life spans.
2. Use your chart to discuss with your teacher and the class the following questions:
   a. Why do some animals live longer than others?
   b. How do scientists determine an average age?
   c. Why has man's life expectancy increased in recent years?
3. Invite some older people to visit your class. Find out why they think they have lived as long as they have.

TEACHER INFORMATION

Animal life spans and cycles vary greatly. Some complete a generation in hours, weeks or months. Others take many years to mature. Life cycles of many animals are focused on continuation of the species. When the reproductive process is complete the animal dies. Natural controls such as predators, available food supplies, and other environmental factors also play a part.

Although many animals have different adaptations and use different methods for survival, man is the only animal to seek consciously to prolong life. For this reason too, domesticated animals tend to live longer than their "cousins" in a natural or wild state.

Students may want to add animals to their life span chart. An encyclopedia or other reference book will usually give life expectancies. Because of variety of reproduction, microscopic organisms are not included in this activity.

Be sure to select the older visitors with care. Conduct a personal interview first and if possible suggest some specific topics to discuss. Children often have more in common with older people than with people in the "middle" years. With care this activity can be a very rich learning experience, and perhaps form lasting bonds.

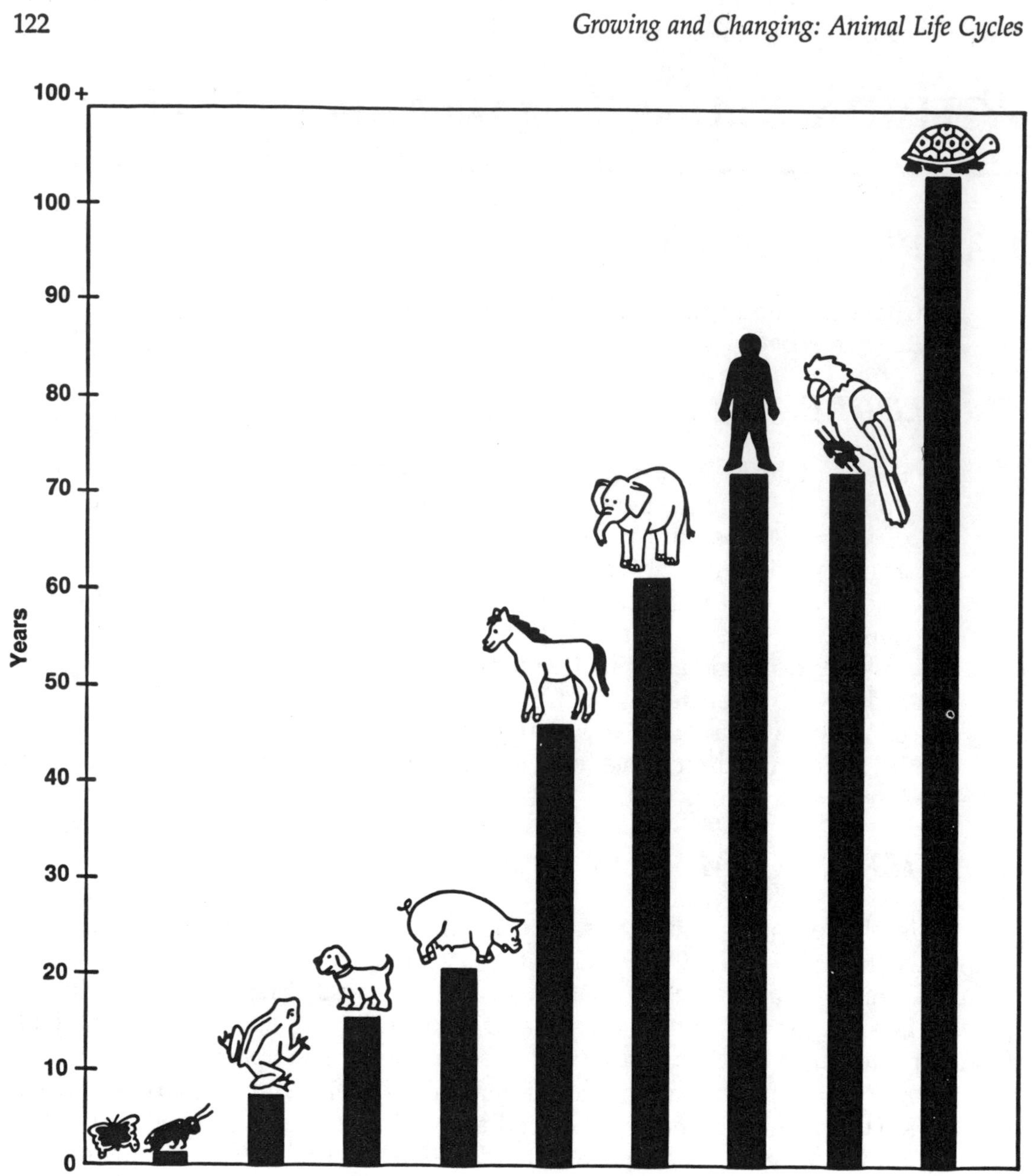

FIGURE 70-1. Animal life spans.

Section 5

ANIMAL ADAPTATIONS

TO THE TEACHER

The following activities are examples of some specific ways animals adapt to their environment. Feathers and birds are used frequently in these activities because they are common even in large cities and suburban areas and many types of adaptations are evident. Feathers collected from wild birds should be carefully examined before use.

The importance of a picture collection is emphasized in the general introduction to this book. This section relies heavily on the use of pictures. Older students might be interested in starting picture collections of their own.

Some household tools used in Activities 75, 76, 77, and 83 are sharp and potentially dangerous. Close teacher supervision is recommended with these activities.

Activity 75 in this section mentions color in our lives. If it seems appropriate,you may want to discuss the use of color in regulatory signs and advertising, just as color is often used in the animal world to advertise and communicate to others.

People are the most adaptable animals on earth. Teachers have frequent need to adapt. As you read and prepare to teach this section, feel free to adapt it in any way you choose in meeting the needs of your students and the conditions of the environment in which they live. Most important, adapt this study according to your needs, special skills, and knowledge.

ACTIVITY 71: What Is Adaptation?

(Small-group discussion)

MATERIALS NEEDED

- Pictures of people living, working, and playing in different climates
- Newsprint
- Pencils

PROCEDURE

1. *Adaptation* in plants and animals refers to the way they can adjust or change to be able to live where and how they do. Many animals and plants can live only in certain places. People have made adaptations so they can live for periods of time almost anywhere. Discuss with your teacher some of the adaptations (adjustments) that have been made by you and others to enable you to be comfortable in your classroom today. List some important ones on the board.
2. Divide into groups of four or five and look at the pictures of ways people adapt to live, work, and play all over the world. Make a list of some ways humans are different from other animals in their ability to adapt.
3. From your pictures, choose the most interesting adaptations and share them with the class.

TEACHER INFORMATION

This can be a very simple activity related to seasons for younger children.

With older students, you may want to relate this activity to your social studies and include historical adaptations, such as those of Eskimos, Indians, pioneers of the past, and primitive people of today. Geography, weather, climate, and food offer almost limitless possibilities for illustrating the human ability to adapt.

For centuries, people have dreamed of exploring and colonizing outer space. Students who have a background in the study of the solar system may choose to explore this topic for enrichment.

Plan to have 50 or more pictures showing homes, food, transportation, recreation, clothing, and everyday activities.

ACTIVITY 72: Why Are Feathers Special?

MATERIALS NEEDED

- Bird flight feathers (one per student)
- Magnifying glass (one per student if possible)

PROCEDURE

1. Use the magnifying glass to examine your feather. Notice that it is made up of a strong shaft or spine, with many small threadlike pieces or *filaments* attached on either side of the spine.
2. Examine the strong shaft. Hold it by the thicker end and carefully bend the opposite end a little bit. What happens when you release the bent end?
3. Carefully separate the threadlike filaments near the middle of your feather. Notice they seem to stick together. Although you probably will not be able to see them with a small magnifying glass, each filament has hundreds of tiny hooks, called *barbules,* that hold the filaments together to make a long, flat surface called a *vane.*
4. Beginning at the shaft, use your fingers to gently stroke the area where you separated the filaments. Can you describe what happens and why? Can you think of man-made devices similar to this?

TEACHER INFORMATION

Before you use feathers, certain precautions should be taken. Be certain they are clean and free of dust and tiny insects such as fleas and lice. Most feathers can be rinsed in warm water (no soap), and when dry, placed in a microwave oven for one to two minutes.

Feathers are remarkable examples of animal adaptation. Man has not been able to produce any material or system to match them, weight-for-weight, for strength, insulation, or air-foil qualities.

The substance of which they are made is called keratin, similar to the material in human fingernails. The tiny barbs on the filaments allow them to be separated and rejoined much as we use zippers or Velcro fasteners.

The filaments of flight feathers lock together in an overlapping pattern to provide extra strength. If you have a large feather and a strong magnifier you may be able to see the overlapping by looking along the shaft from a nearly horizontal angle. Many of the following activities use birds as examples of adaptation. Some general bird characteristics are introduced in Activities 46 and 54 in the section on "Animals."

ACTIVITY 73: What Purposes Can Feathers Serve?

MATERIALS NEEDED

- Flight feathers
- Downy feathers
- Magnifying glasses (one per student)
- Pictures of penguins in natural habitat
- Pictures of people in heavy winter clothing
- Man-made insulation from a coat, sleeping bag, or ski parka

PROCEDURE

1. Use your magnifying glass to compare your flight feather from Activity 72 with your new downy feather.
2. Locate shaft and filaments. How are they alike? How are they different?
3. Can you think of reasons birds have downy feathers?
4. Because they are warm-blooded and must use energy to produce heat, both birds and people keep warm by using protective coverings. Use your magnifying glass to examine man-made synthetic materials. Compare these materials with your downy feathers. How are they alike? How are they different?
5. Compare the penguins and the people dressed in winter clothing. Although penguins do not fly, their outer feathers serve as a protective waterproof covering for an inner lining of downy feathers. They are the only warm-blooded animals that can survive year-round in the severe cold of the Antarctic. Without extra heat and shelter, humans would perish in a few hours or days. The warmest coats people can make are made from the downy feathers of birds.

TEACHER INFORMATION

Down from the eider duck has long been prized as the best insulation for coats and parkas. Fur and wool are inferior if heat retention alone is desired. In the extreme temperatures of outer space, astronauts have bulky heaters and air conditioners built into their space suits.

It is important for students to understand that downy feathers and penguins are only *one* example of adaptation to a particular environment. It is equally important to point out that man continually improves his ability to adapt through the use of intelligence and technology.

ACTIVITY 74: Can You Make a Feather?

MATERIALS NEEDED

- Feathers from Activities 72 and 73
- Light-colored construction paper
- Thin plastic drinking straw (not hinged)
- Strong glue
- Scissors
- Pencil

PROCEDURE

1. Place your flight feather on a piece of construction paper. Outline it with your pencil.
2. Use your scissors to cut the outline of the feather from the construction paper.
3. Cut the straw so it is approximately the same length as the shaft of your flight feather.
4. Glue the straw on the paper feather in approximately the same position as the shaft of the real feather.
5. Compare the feather you have constructed with the real feather. Test the weight, strength, *resiliency* (ability to bounce back when bent). What happens if you get it wet? Torn?
6. Would a bird like to have feathers like the one you made?

TEACHER INFORMATION

True feathers are unique to birds. Although all birds have feathers, many do not fly. The oldest known fossil of a bird, archaeopteryx, had feathers, but probably could not fly because it appears to have had solid bones that would have made it too heavy. Small and apparently simple as it appears, the unique design of the feather cannot be improved by modern technology.

Usually birds shed and replace their feathers annually through a process called molting. Animals with hair and fur (including humans) also shed and replace their covering on a regular basis.

Birds regularly use their beaks to groom and repair their feathers (zip them back together) and keep them in place so they function properly. The word *preen* is used to describe this instinctive, necessary behavior in birds and is sometimes applied to vain humans. Many birds have an oil gland located near their tails that secretes a natural protective substance that is applied to their feathers with their beaks.

The feathers and down of many birds attract fleas and lice. Most birds are attracted to water. A large clear plastic bowl filled with water may attract as many birds as a feeder. Be sure the bowl is suspended or placed in an area that is cat-proof.

ACTIVITY 75: How Is Color Used by Living Things?

(Small groups)

MATERIALS NEEDED

- Pictures of brightly colored male and drab female common birds (pheasants or robins)
- Pictures of brightly colored and drab fish
- Pictures of humans dressed in bright colors and camouflage
- Pictures of butterflies and insects that resemble a leaf or bark
- Pictures of doe and fawn

PROCEDURE

1. Color is important to many living things. Look at the pictures. Classify them according to bright and drab colors.
2. In many species of animals the female is not as brightly colored as the male. Try to match the male and female of each species. Some male animals use bright colors to attract the female. The drab colors of the female protect her while she is caring for her young by making her less noticeable.
3. Some animals and plants on both land and sea are adapted to protecting themselves by resembling their natural surroundings. Can you find animals that use this method, which is called camouflage? One of the best examples of camouflage is the chameleon, a lizard that changes its skin color to resemble its surroundings. Some birds and fur-bearing animals (for example, rabbits) change color with the seasons.
4. Look at the pictures of humans. How have they used color? Why?

TEACHER INFORMATION

Color, or lack of it, plays such as important part in nature that it could occupy a portion of almost any area of study. People appear to be unique in using color for purely aesthetic purposes. Because the basic physical appearance of humans is quite drab, they have again adapted widely from nature. From earliest times feathers, furs, jewelry, and dyes have all been used as adornments by people, and from prehistoric times people have also used color in art.

Older students may be interested in studies of the psychological effects of color on animal and human behavior. Consult your encyclopedia for additional information. Do bulls react to the color red? No, but hummingbirds do. They love it.

ACTIVITY 76: How Do Birds Adapt to Eating?

(Teacher-supervised total-group activity)

MATERIALS NEEDED

- Variety of pictures of birds with many kinds of beaks (name of food they eat written on back)
- Household utensils and tools including scissors, two-tined fork, linoleum cutter, sieve, knives (serrated and plain), wooden spoons, nutcracker, salad tongs (hinged at end), tweezers, scissor tongs, tea ball, medicine dropper, needle-nosed and standard pliers, hammer, ice pick or punch, chopsticks, grater, ear syringe

PROCEDURE

1. Arrange the pictures of birds on a large table. Carefully study their beaks. What type of food do you think their beaks are best designed for?
2. Look at the household tools. Do any of them remind you of birds' beaks? Be sure you know how each one works.
3. Examine each household implement and put it next to the bird whose beak it most closely resembles. Can you think of other objects around your home or school that work the same as birds' beaks?
4. Test yourself by turning the bird pictures over. On the back is a list of foods they eat.

TEACHER INFORMATION

Since some implements are sharp, this activity should be carefully supervised. (To avoid the risk of having the sharp instruments, pictures of them could be used instead of the instruments themselves.) Birds do not have teeth, but some of their beaks or bills are highly specialized, depending on the food they usually eat. The terms *beak* and *bill* are generally used interchangeably. Some ornithologists reserve the term *beak* for the sharp, hooked beaks of hawks, eagles, owls, and other birds of prey.

The most common beaks among land birds are varieties of the short, strong seed eaters. If you have a canary or parakeet visiting in your classroom notice that before they eat some types of seeds, they use their beaks to split and remove the outer covering or husk.

Ducks and many other birds with flat bills often use them in water as strainers or sieves to draw in plants, snails, and small aquatic animals. The varieties, range, and eating habits of ducks make them easily available for study throughout the moderate climates of the world.

Some birds (for example, grebes and cormorants) with long scissorlike bills catch small fish by diving and swimming under the water for short periods of time. Other fishing birds (herons and egrets) stand on long legs and fish near the shore.

Marsh and shore birds (for example, sandpipers) have long, often curved bills to catch shrimp and other crustaceans in marshlands and at the seashore.

Some birds (terns, pelicans, kingfishers) circle in the air above the water and dive under to catch fish with their scissorlike bills.

The bills of woodpeckers are mostly long and sharp and have built-in shock absorbers to chisel out tree bark insects and construct nest holes. Some woodpeckers also eat acorns and other nuts, often drilling holes in trees as storage bins for their food.

Some birds (pigeons, jays, gulls) eat a wide variety of foods and may become bothersome intruders at picnics and campsites.

The tiny hummingbird uses its very long bill much like a medicine dropper to draw in nectar and pollen. Despite its small size, the hummingbird is an excellent flyer and can hover, dart at high speeds, and even fly backward. Hummingbirds are highly territorial, somewhat aggressive, and usually friendly and curious.

Adaptations of birds are almost unlimited. Probably the most feared and respected are predatory birds (eagles, hawks, ospreys, falcons, owls) with sharp, hooked beaks and powerful claws. Although of different scientific classifications, these birds, often referred to as *raptors*, have, in addition to strong, sharp beaks and claws, superior flying ability and excellent eyesight.

ACTIVITY 77: What Are Some Other Eating Adaptations?

(Teacher-supervised total-group activity)

MATERIALS NEEDED

- Household utensils from Activity 76, plus *picture* of hypodermic needle
- Pictures of animals with mouths open showing mouth parts. Write the name of principal food they eat on the back. Try to include animals that live almost entirely in the water such as: shark and whale, squid, octopus; other types of animals such as: reptiles (crocodile, alligator, rattlesnake, constrictor, lizard with long tongue;) common mammals—including rodents, dogs, wolves, bears, sheep, cows, horses, cats (large and small), bats, shrews, beavers, and deer family; and representative dinosaur pictures. Try also to include enlarged insect mouthparts (mosquito, flea, ant, bee, grasshopper, praying mantis) and enlarged spider mouths.

PROCEDURE

1. In Activity 76 we found that birds' beaks are often adapted to the kinds of foods they eat. Study the pictures on the table and try to arrange or classify them into different groups whose open mouths look similar.
2. Look at the household utensils. Try to find some that look like (resemble) the mouthparts in the animal pictures. Put the utensils near the pictures. You may find some animal mouthparts that have several utensils in them.
3. Look at the pictures of the dinosaurs. The largest known dinosaur, *Brontosaurus*, had a long neck and small mouth. Compare it with the picture of *Tyrannosaurus Rex*. Compare the mouths of these giants from the past with the mouthparts of a spider or praying mantis. Can you tell how large the animals are by looking at their mouthparts?
4. One way of classifying animals is by the kind of food they eat. Some animals eat plants, some prefer meat, and some eat both. Arrange your animal pictures into these three groups according to the food they eat. Turn the pictures over and see if you were right. What were the clues that helped you decide?

TEACHER INFORMATION

As with birds, mouthparts of different species of animals vary greatly. Sharp, curved, tearing teeth (similar to eagles' beaks) usually indicate meat eaters. Cutting and grinding teeth (or the absence of teeth) usually suggest plant eaters. Animals that eat both usually have some combination of tearing, grinding, and cutting teeth.

There appears to be no simple relationship between the size of animals and the food they eat. Quantity and availability of food seems more significant. The largest of the dinosaurs were plant eaters and so are elephants today.

When the giant lizards ruled the earth, mammals were very small and not plentiful. Today, some mammals are very big. Some species of whales (straining tons of tiny sea animals through their sievelike mouths) are three to four times larger than *any* other animal that has *ever* lived.

The traditional terms *herbivore* (plant eaters), *carnivore* (meat eaters), and *omnivore* (both plants and meat) have not been used in this section. Bears, pigs, people, birds, bats, and many other animals feed in such a variety of ways that the terms may become quite confusing. Your students may also have house pets (cats and dogs) that are basically meat eaters but have become so domesticated that they will eat anything you do, even french fries!

ACTIVITY 78: What Are Human Teeth Like?

MATERIALS NEEDED

- Dentures
- Tooth model
- Ear syringe
- Picture of mother dog or pig nursing her young
- Picture of typical six- or seven-year-old child with front teeth missing
- Picture of horse or mule showing its teeth

PROCEDURE

1. What does the saying "Don't look a gift horse in the mouth" mean? Look at the picture of the smiling horse. How does it make you feel? If you were a horse dealer you might react in a different way. Horses eat hay, grass, and some other plants. Over the years their teeth gradually wear down. Looking in a horse's mouth is one way to tell its age. Until recently, looking in a human's mouth might tell the same story. Why is this no longer true?
2. In Activities 76 and 77 you learned some ways mouthparts are important to animals in order to get and eat food. Many of the mouthparts were developed for one kind of food. A few had parts for eating a variety of foods. Examine the dentures (false teeth). They are very similar to the ones you have in your mouth. Find teeth that look like the beaks, bills, and specialized mouthparts of other animals. From looking at their teeth can you tell what humans eat?
3. Baby mammals, including humans, usually don't have teeth when they are born. They get their food in the form of milk by sucking on the soft nipples of their mother. Examine the ear syringe. Put some water in your hand and draw it up. You can draw milk or water into it without damaging the surface it touches. Mammal babies seem to be born with an instinct for sucking.
4. Sometimes at birth but usually after a few months, human babies begin to develop teeth. The first ones are small but allow babies to begin eating soft food other than liquids. These first teeth are temporary and usually fall out. Do you have temporary teeth in your mouth? Find a picture of young children who are losing their temporary teeth.
5. Look at the set of false teeth. By the time you are a young adult (16 to 20 years old) you will have a permanent set of teeth, usually 28 to 32 in number. With care, they will serve you well for the rest of your life.
6. If you haven't figured out what "Don't look a gift horse in the mouth" means, discuss it with your teacher.

TEACHER INFORMATION

Human teeth are also studied in Activities 144 and 145 in the section "Health and Nutrition." Activity 144 develops tooth structure; Activity 145 emphasizes tooth care. Activities 78 and 79 in this section use a somewhat different approach. You may prefer to integrate information from both sections and present them at one time.

Dentures, real teeth, charts, and models may be obtained from a local dentist. The American Dental Association also provides pamphlets and information about teeth and their care.

"Don't look a gift horse in the mouth" may have originated at the time when horses were common in everyday life as cars are today. If someone gave you a horse as a gift and you looked in its mouth to see how old it was it would be as rude as to look for a price tag or brand name on an item received as a gift. Today we often use the expression to describe someone who has had good (unexpected) fortune or received a gift and looks for an underlying or selfish motive behind the gift.

The human is very adaptable and its teeth and mouth reflect this pattern. We have fairly good cutting teeth (not as good as those of a beaver), and grinding teeth (but not nearly as good as those of a cow). Our mouths can be shaped for sucking but not nearly as efficiently as that of a hummingbird, mosquito, spider, leech, or lamprey.

Where food is concerned, the size of our mouths and the limited operations of our jaws keep us from being the best at anything, but because they are so generalized we do pretty well at everything.

Human age can sometimes be determined by counting permanent teeth (6- and 12-year molars). Until recently the presence of dentures was associated with old age.

ACTIVITY 79: Why Are Healthy Teeth Important to You?

(Teacher-directed small-group activity)

MATERIALS NEEDED

- Pictures of smiling people of all ages and different nationalities (try to include Eskimos and American Indians) with healthy teeth
- Dentist or dental hygienist as resource person
- Pictures of cave men or primitive people (eating if possible)
- Newsprint
- Pencils

PROCEDURE

1. As you listen to the dentist or hygienist, use your pencil and paper to list as many reasons as you can why healthy teeth are important and how you can keep your teeth healthy.
2. Look at the pictures of different people with healthy teeth. Some work very hard to keep them healthy and clean. Other groups do almost nothing special. Can you think of reasons why? Before you do step 3, write down any reasons you can think of.
3. Look at the pictures of the cave men and primitive people. Scientists have found evidence that these cave men and primitive people ate their food almost as they found it. They gnawed bones, chewed skins, and ate vegetables and fruit without boiling, softening, or tenderizing them. Because their teeth worked hard and were scraped, scratched, and brushed by tough foods, tooth and gum problems were rare. Broken teeth were common. Was this one of the reasons you listed on your paper?
4. Some highly developed and civilized people such as American Indians and Eskimos have very healthy teeth when they follow the customs and eating habits of their culture. Before white people came, most American Indians and Eskimos had very healthy teeth. Now many do not. Can you think of reasons why?
5. Good dental hygiene (brushing and flossing) is very important for everyone, but even if we try very hard, some people will have better teeth than others. This is also true of eyesight, hearing, and many other things. People adapt to different problems in different ways. Following is a list of some reasons for good and poor teeth and gums. Some you can change and some you cannot. Make list of the things you can change and tell how.
   a. Tooth decay from eating and drinking foods with sugar.
   b. Tooth decay from not brushing after meals and flossing once a day.

c. Tooth decay due to lack of fluoride in your drinking water.
d. Crooked teeth
e. Overbite
f. Sore gums
g. Color of teeth
h. Tooth problems of parents and grandparents
i. Eating foods that are hard to chew.

6. Discuss the list in step 5 with your teacher and your dentist.

TEACHER INFORMATION

Although teeth have not changed significantly in the history of modern people, their lifestyle and eating habits have. Good dental care, hygiene, and nutrition (sugar control) are people's ways of adapting their teeth to constant change.

Different cultures and varying lifestyles influence the amount and type of care human teeth require. Heredity must also be considered when evaluating tooth development and care.

ACTIVITY 80: How Do Animals Move?

(Small-group activity)

MATERIALS NEEDED

- Pictures of animals that move in different ways (swimming, walking, hopping, gliding, crawling, flying, climbing) and of animals that don't move, such as barnacles
- Live animals in aquarium, cages, and terrarium
- Paper
- Pencils

PROCEDURE

1. Look at the live animals and the pictures of animals in your room. Can you tell how each one moves about? Can you find any that don't seem to move?
2. Write the names of the animals on the left side of your paper. Next to their names, write the ways they move (remember some move in different ways). Try to think of more than one word to describe their movement, for example: fish swim, but they also wiggle and swish; snakes crawl, but they also slither and glide; cats walk, run, crawl, creep, climb, stalk, and pounce. How many words can you find for each animal?
3. Think of the ways you move. Look at your list of animal movements. Underline the ones you can imitate. Are there any movements you can do as well or better than some other animal (hint: pick up a pencil or tie a shoelace)?
4. Have a class discussion about why you think animals move as they do.

TEACHER INFORMATION

Younger students may enjoy trying to imitate the movements of animals. Older students can play animal movement charades, in which they guess the animal being portrayed. Animal movements may also inspire creative dancing. Some cultures imitate animal movements with costumes, music, and dance.

Animal movements are most often based on their environment (water, land, or both), method of obtaining food, reproduction, and defense or protection. Some animals seem to dance or play just for fun.

When size and proportions are taken into consideration, animals that specialize in a particular type of movement can perform that movement far better and more efficiently than people (even remaining still). Because of the opposed thumb, the only movement people can make better than any other animal is grasping. This has enabled people to be the best "tool users" and to create many technical and mechanical adaptations to compete with the specialized movements of other animals.

Don't forget the many mechanical robots that are popular as children's toys. You might plan a "robot day" when everyone is invited to bring his or her favorite doll or mechanical animal toy and explain how it works. No matter how well designed, mechanical toys cannot perform animal movements nearly as well as the animal itself.

ACTIVITY 81: How Do Some Animals Communicate?

(Partners or small-group activity)

MATERIALS NEEDED

- Paper
- Pencil

PROCEDURE

1. We usually think of communicating as ways of giving or receiving information or conveying emotions and feelings by talking and listening. Listening and watching (movies, radio, television) or drawing and looking at pictures are also usual ways man communicates. Can you think of any other ways?
2. In step 1, did you add reading, which you are doing now? Did you remember gestures, too? On your paper, write as many ways as you can think of that people communicate.
3. Scientists believe that many animals communicate in different ways, but only man has been able to create a vocabulary of many words to communicate ideas. If you own or know a household pet, discuss the ways it communicates with you without using words. Does it seem to understand what you say? Write down as many of these ways as you can.
4. With partners or in small groups, pretend none of you can hear or speak. Without drawing pictures or making any sound, take turns describing your favorite game or toy.
5. You probably found your best way of communicating was with gestures. Many people who cannot speak or hear learn gestures called sign language or "signing" to communicate. List on your paper some animals that use gestures or signals to communicate (hint: How does a dog show it is friendly?).

TEACHER INFORMATION

In addition to the opposed thumb, the abilities to create and use oral language and pictures seem to be unique to people. There seems to be little doubt that other higher animals such as apes can reason and think, but only people are known to have developed a complex language with which to express ideas.

Apes and chimpanzees have been successfully trained to use symbols and sign language (see *National Geographic*, January 1985).

Most other animals seem to use communication instinctively to express warning, fear, anger, recognition, and food gathering, or as mating calls and signals. The famous "bee dance" is described in your encyclopedia. Ants also have methods for communicating the location of food. Most insect communities provide studies of specilization and ways animals within a species are adapted to specialized roles. Highly intelligent sea mammals, such as porpoises and whales, seem to have complex communication systems that scientists have, so far, been unable to decode.

Language has enabled people to use the environment to adapt, create, and express as no other animal can. In recent years, the whole new universal language of computers has been created. As people continue to adapt it to every phase of life the implications seem immeasurable. For sheer joy of expression, however, many still believe that the infinitely varied song of the humpback whale has no equal.

ACTIVITY 82: How Do Seasons Affect Animal Adaptations?

MATERIALS NEEDED

- Pictures of butterfly, migratory bird (Arctic tern, Canada goose) local year-round bird, squirrel, rabbit, earthworm, snake, bear, deer, whale, and pictures of people from Activity 71
- Map of the world
- Globe
- Cards labeled "hibernate," "migrate," "adapt"

PROCEDURE

1. Place the three cards at the top of the table. Discuss the meaning of each word with your teacher.
2. Look at the map and globe of the earth. Usually there is less seasonal change near the middle, or the *equator.* As you travel north or south toward the poles, seasonal changes become greater. Find the place where you live. How far are you from the equator, and from the North Pole or South Pole?
3. Oceans and other large bodies of water affect land temperature. Water does not change temperature as rapidly as land. If you live near an ocean, changes in seasons may not be as great. Does a large body of water affect your climate? How?
4. Study the pictures of animals. They are all living but not all are adults. See how many pictures you can match with the three cards to show how these animals adapt to seasonal change.
5. With your teacher and other members of the group, discuss each picture and put it in one of the three groups. Why do you think animals migrate or stay and adapt to seasonal change?

TEACHER INFORMATION

Animals have different ways of adapting to seasonal change, mostly due to the availability of food. Some hibernate in a completely resting stage. Some animals, such as bears, increase their rest and reduce their level of activity but do not completely hibernate.

Many animals remain active in the same area year-round and utilize whatever food is available. This is more often the case in mild and warm climates.

Some animals also migrate and gather for breeding purposes. They usually select isolated, protected areas free of their natural predators. Climate and season are also factors. Activity 83 develops an unusual example of animal adaptation to seasons.

Animal migration is a fascinating area of study. You may find high student interest and extend your study beyond the information introduced in this and the following activity.

In past times caribous, buffaloes, and reindeer existed in great numbers. As these animals migrated with seasonal change, men who depended on them for their existence (food and clothing) also migrated.

ACTIVITY 83: How Do Animals Adapt to Seasonal Change?

(Small-group activity)

MATERIALS NEEDED

- Pictures, map and globe from Activity 82
- Picture of Arctic tern (if not in above collection)
- String
- Masking tape
- Scissors

PROCEDURE

1. Find the picture of the Arctic tern. It is most famous as a world traveler. Locate Greenland on the globe. Many baby terns are hatched here in late June or early July. Place a piece of masking tape here.
2. Within a few weeks the terns are ready to begin their migration. Use the globe, a 50-cm. (20-in.) piece of string, and a piece of masking tape to mark their journey.
3. After leaving Greenland, they fly to the west coast of Europe. Stretch your string and tape it from Greenland to the west coast of France, Spain, or Portugal.
4. From this point they fly down the west coast of Africa. Stretch your string and tape it to at least two points near the top and bottom of Africa's west coast.
5. From a point near the southwest part of Africa they make a long journey over the south Atlantic and the Antarctic Ocean to the crusted ice of Antarctica. Stretch your string across the oceans to Antarctica and tape it there. Because the seasons are opposite in the Northern and Southern *Hemispheres* of our earth, it is summer in the south when it is winter in the north.
6. Terns spend several months in Antarctica, during which they may circle the entire continent. Make a circle with your string around Antarctica.
7. Look at the string path from Greenland to Antarctica on your globe. This marks half the distance a tern flies each year. By May, winter is begining in Antarctica and the tern returns to Greenland over approximately the same route. Mark or cut the string where the tern's journey ends in Antarctica. Remove the string from the globe. Measure the length of the string and cut another piece twice as long. This represents (stands for) the total distance a tern may travel each year.
8. Wrap the longer piece of string around the equator or middle of the globe. Then wrap it around the globe, crossing both the North Pole and the South Pole. Can you see why we call terns world travelers?
9. Other animals in the pictures travel shorter distances, and for different reasons. Some birds in the United States and Europe migrate east and

west rather than north and south. Why? Choose the picture of an animal that migrates. Use your encyclopedia and books from the media center to find out as much as you can about migrating animals. Share your investigation with the class.

TEACHER INFORMATION

Availability of globes and the ages of the students may limit the size of groups. Try to borrow as many globes as possible so students can work out the problems themselves.

Basic skills in map and globe reading are required. These skills are discussed in *Earth Sciences* (Book III of this *Library).* You may want to review the activities on map and globe skills and seasonal change found in that book's Section 7, "Above the Earth" and Section 8, "Beyond the Earth."

Some birds in Europe, North America and other parts of the world migrate east and west rather than north and south because they "summer" inland and "winter" near the coast where it is warmer. Ornithologists are not certain how birds navigate or "home." Theories include the use of stars, the earth's magnetism, landmarks, and combinations of these. Their ability to find a specific location on the earth far exceeds that of man and they do it even in great numbers without the help of traffic controllers!

ACTIVITY 84: Can You Design a Better Bird?

(Teacher-assisted small groups or partners)

MATERIALS NEEDED

- Pictures of birds in flight, on land, on water
- Paper and straw feathers constructed in Activity 74
- Balloons of different sizes and shapes
- Small Styrofoam balls
- Plastic drinking straws
- Colored construction paper
- Cloth or nylon stockings
- String
- Strong white glue
- Pencils and crayons
- Scissors
- Newsprint
- Wire coat hangers (optional)

PROCEDURE

1. Choose a picture of a bird you like. On the table are materials to construct a bird of your own.
2. With your partner or group, make a plan for a bird made of balloons, Styrofoam balls, paper feathers, straws for legs and feet, and a household utensil for a beak or bill.
3. On a sheet of newsprint draw and color a picture of what you think your bird will look like. This will be your construction plan. Be sure to decide whether your bird will be flying, standing, walking, sitting (perching), or swimming.
4. Balloons and Styrofoam balls will make good bodies and heads. Straws can be used for feather parts, legs, and feet. If a household utensil is not satisfactory, you may want to make the beak or bill out of something else. Remember, you will need to make many feathers of different colors. You may want to share your plan with your teacher before you begin.
5. After your bird is finished, share it with other students in the class.
6. Give your bird a name. On a piece of paper write its name, what it eats, where it lives, and why it is special.
7. Make an "aviary" for all the birds in your class. Invite visitors to see it.

TEACHER INFORMATION

This activity could be used at the end of your study of animal adaptation. You will need many paper feathers. If different sizes and colors are cut out from a pattern in advance, much time can be saved. Students may need help in forming the tail, as its shape and construction are often not as obvious as they are for wings.

Coat hangers with the tops cut off can be used as rigid supports for the wings of flying birds. If you use coat-hanger wire, be sure to cover the sharp ends with tape (saves balloons).

No matter how tightly you seal them, balloons will gradually lose air. Inflate them last and plan to have your birds on display for only a few days. If you care to invest the time, the balloons can be covered with papier-mâché or glue-dipped strips of newspaper to increase the longevity of your birds.

Section 6

BODY STRUCTURE

TO THE TEACHER

This section provides opportunities for students to learn about themselves. Activities involve both the muscular and skeletal systems of the body. Students can discover and learn a great deal about their own body structure by studying the bones and muscles of animals. These parts are usually available from the local meat market. Hair, nails, skin, and lung capacity are also dealt with in the activities of this section.

The teacher should invite resource people into the classroom at appropriate times to enrich the experience. Along with the section "Health and Nutrition" and "The Five Senses," the study of this section provides many excellent opportunities to explore the world of work with respect to the health services occupations and professions.

In a study of the structure of the body, the handicapped should be studied and recognized as normal people for whom certain abilities are limited. There is such a broad range of ability among the nonhandicapped, and such a broad range among the handicapped, that it is sometimes difficult to distinguish between the two. It is hoped that in the study of this section, physical differences will be recognized and treated as normal. We are all different. Being different is normal. It should be noted and stressed, however, that we are more alike than different.

We suggest that the teacher scan all activities in the section before beginning its use in the classroom, taking note of materials required. This will aid in making necessary advance preparations.

ACTIVITY 85: How Big Is the Average Person Your Age?

MATERIALS NEEDED

- Group of students your age
- Pencils
- Butcher paper
- Metric measuring tape
- Masking tape

PROCEDURE

1. Have a friend measure and record your height, your arm span, your hand span, and the length of your foot (shoes off).
2. Measure your friend in the same way.
3. Compare your measurements with those of your friend.
4. Compare your arm span with your height.
5. Figure the average height, arm span, hand span, and foot length in your class.
6. Fasten a piece of butcher paper to the wall and mark the average height for your class. Draw on it a hand of average size and a foot of average size.
7. Fasten two more pieces of butcher paper to the wall and record the tallest height and the shortest height, and the longest and shortest of the other measurements. Do not write names on these.
8. Examine and compare the various measurements of the class. Who seems to be closest to average in every measurement? Where do you fit in with each of the measurements taken?

TEACHER INFORMATION

Students usually enjoy comparing their size with that of others. If you avoid including waist size or weight in the measurements, the risk of offending someone should be very slight. It might be necessary to review the process of computing averages before the students do step 5. Students too young to compute averages can still measure each other and compare. They can talk about size ranges and the sizes most students this age seem to be. Students should be taught that it is okay to be big or small. Such people are normal. We are all different and grow at different rates.

Here are some suggested extensions to this activity:

1. Have a group discussion about the ways people are different—the characteristics we notice in distinguishing one person from another.
2. Assign students in pairs and have each one draw an outline of the other on butcher paper.

3. After the average measurements have been computed for the class, assign a group to draw "Mr. Average" or "Miss Average" on butcher paper.
4. Try to find out how various forms of measuring were developed: inch, hand, foot, yard, fathom, cord, acre, cubit, meter. Which of these originated from a measure of the human body?

Note: At step 4, children will usually find their arm span to be slightly greater than their height, although arm span from midfinger to midfinger usually approximates height.

ACTIVITY 86: How Do the Body Systems Work Together?

MATERIALS NEEDED

- Bicycle
- Picture of body showing internal organs

PROCEDURE

1. Examine the bicycle and explain how it works. For each part that moves, tell what makes it move. How do the different parts depend on each other?
2. Look at your body and tell what parts you think depend on other parts.
3. Bend down and pick something off the floor. How did your hand depend on your arm, your arm depend on your shoulders, your shoulders depend on your back, your back depend on your legs, and your legs depend on your feet?
4. Compare the way your body parts depend on each other with the way the bicycle parts depend on each other.
5. Look at the picture of the body's internal organs. Can you add to your explanation of parts of the body that depend on each other?

TEACHER INFORMATION

Every part of the body depends in some way on many, many other parts of the body. Point out that the control center for all of the body parts is the brain. The "brain" of the bicycle is the person riding it.

Consider making similar comparisons with other machinery in addition to the bicycle, such as a typewriter, pencil sharpener, door latch, and so forth. If someone in the class knows something about automobiles, this person could be asked to explain some of its interdependent systems. This might be a good time to invite a resource person to the classroom.

ACTIVITY 87: What Is Your Lung Capacity?

MATERIALS NEEDED

- Gallon bottle
- Sink or large pan
- Tubing
- Measuring cup
- Water
- Masking tape

PROCEDURE

1. A bottle with a small opening, such as a cider or vinegar bottle, will work best.
2. Put about 5 cm. (2 in.) of water in the sink.
3. Fill the bottle completely with water.
4. Cover the top of the bottle, turn it over, and stand it upside down in the sink with the opening in the water. When you uncover the opening, no air should enter the jug.
5. Tip the bottle slightly to the side and insert one end of the tube into the opening of the bottle.
6. Have your partner hold the bottle upright while you do steps 7 and 8.
7. Take a deep breath and blow through the tube, emptying the air from your lungs as completely as you can into the bottle.
8. Mark the water level on the bottle with a piece of tape.
9. Empty the bottle, turn it right side up, and use the measuring cup to measure the amount of water required to fill it up to your mark. This is

FIGURE 87-1. Student measuring his lung capacity.

the amount of air you blew into the bottle. It is the vital capacity of your lungs. In addition, you have about 20 percent of residual or "dead" air still left in your lungs which you cannot expel.

10. Repeat the activity for your partner and others if they wish. Compare and compute the average lung capacity of those who participate.

TEACHER INFORMATION

This is a valuable activity for practice in measuring a volume of air and in getting acquainted with the body. You might have someone in the class investigate the amount and types of physical exercise in which each group member participates. Then search for correlations between lung capacity and exercise. Someone else could measure the height of each person and search for a correlation between height and lung capacity.

The lung capacity measured in this activity is called the vital lung capacity. It is less than the total lung capacity because some air remains in the lungs after exhaling as much as possible.

ACTIVITY 88: How Fast Do Your Nails Grow?

MATERIALS NEEDED

- Nail polish
- Ruler
- Paper and pencil

PROCEDURE

1. Put a tiny spot of nail polish next to the cuticles of one fingernail and one toenail. Let it dry. Plan to leave it there for several weeks.
2. Check the nail polish each day. If it begins to wear away, put another spot of polish on, but be sure to put the new spot exactly on top of the old.
3. Each week, measure the distance from the cuticle to the spot of polish and record it. Do this until the spot of polish grows out to the point that you cut it off when you clip your nails.
4. What was the average weekly growth of your fingernail? Your toenail?
5. Did either your fingernail or toenail grow faster than the other? If so, which one, and how much faster?

TEACHER INFORMATION

A line could be scratched into the nail with a nail file and the polish applied on the scratch. This will increase the life of the polish on the nail and help to assure accurate replacement if it does wear off.

If several students are involved in this activity, have someone compile the results and make a graph. If one person makes the graph, each student could plot his or her own results. From the group results, with or without the graph, average fingernail and toenail growth rates could be computed for the class. Perhaps some could carry the research a bit further and include other members of their family, thus finding out whether age seems to be a factor in nail growth rate. Comparing results between families will provide indicators of heredity as a factor.

Fingernails normally grow about three times as fast as toenails.

ACTIVITY 89: How Do Fingerprints Compare?

MATERIALS NEEDED

- Ink stamp pad
- Paper

PROCEDURE

1. Have each participant place the tip of the right forefinger on the ink pad with a slight right-to-left rolling motion.
2. Immediately after applying ink to the finger, place it on the paper, using the same rolling motion.
3. Examine the fingerprints.
4. How are they alike?
5. How are they different?
6. Do you notice any patterns that are similar in some of the fingerprints?
7. Do you see any two fingerprints that are exactly alike?

TEACHER INFORMATION

Some participants might want to take the prints of all their fingers and study the similarities and differences. They will find that no two fingerprints are exactly alike, not even from two fingers of the same person.

Discuss with the group why fingerprints or footprints are used for identification in some important documents, such as birth records and police records. See also Activities 45 and 46 in Section 3, "Animals."

ACTIVITY 90: What Does Hair Look Like Under a Microscope?

(Teacher-supervised activity or teacher demonstration)

MATERIALS NEEDED

- Microscope
- Razor blade

PROCEDURE

1. Remove a hair from your head.
2. Put the hair under the microscope and examine it.
3. Describe the hair. What do you notice about the hair by looking through the microscope that you cannot see without the microscope?
4. Ask your teacher to slice through the hair diagonally with the razor blade.
5. Now examine the hair under the microscope again. Notice particularly the diagonally cut end. Do you see layers? How many? Where does the color seem to be darkest?
6. Trade hairs with someone else and compare the structure and coloring with yours. What similarities do you see? What differences?
7. Remove a hair from your eyebrow and an eyelash and examine these under the microscope. How do they compare with the hair from your head?

TEACHER INFORMATION

The use of the razor blade in this activity must be carefully supervised. It is best that you operate the razor blade yourself, for obvious safety reasons.

Students should be able to see three layers in the diagonally cut hair. The middle layer is the one that contains the pigment, providing color. This, of course, will be more evident with hairs of darker color. As people grow older, the pigment sometimes disappears and the hair turns white. Students should also notice other qualities, such as that some hair is coarse and some is fine.

See also Activity 41 in Section 3, "Animals."

ACTIVITY 91: How Does Our Skin Protect Us?

MATERIALS NEEDED

- Four apples
- Straight pin
- Alcohol
- Four sheets of paper
- Cotton swab or paper towel
- Pencil
- Writing paper

PROCEDURE

1. Lay out the four sheets of paper on a table or a shelf. Label the papers A, B, C, and D. Carefully wash your hands.
2. Wash the four apples and place one apple on each paper. Let the labels on the papers identify the apples. You will notice that apple D remains untouched from this point on.
3. Use straight pins to puncture four holes in apple B and four holes in apple C.
4. Have someone with unwashed hands rub his or her hands around on apples A, B, and C, including the punctured areas of apples B and C.
5. Use the cotton swab or paper towel to apply rubbing alcohol to the punctured areas of apple C.
6. Leave all four apples in place, unhandled, for seven days. Each day, examine (but don't touch) the apples and write a description of any changes you observe.
7. After seven days, compare the four apples. Consider the possible effect of the rubbing with dirty hands, the punctures, and the applications of alcohol.
8. Why did we use apple D?
9. Compare the apple skin to your own skin. What can you say about what you observed with the apples?

TEACHER INFORMATION

Apple D was washed and left untouched to serve as a control. This is an opportunity to stress the importance of the use of controls in many experiments, and the ease of including a control.

The apple skin protects the apple in much the same way as our own skin protects us. This should be evident in comparing apples A and D. When foreign substances do penetrate the skin, such as through an open wound, the importance of using an antiseptic should be evident by comparing apples B and C.

ACTIVITY 92: How Does the Skin Help Regulate Body Temperatures?

MATERIALS NEEDED

- Water

PROCEDURE

1. Wet your finger and blow on it. How does it feel?
2. Wet a spot on your arm and blow on it. How does it feel?
3. Think about when you first get out of a shower or a bathtub. How do you feel while you are wet? Do you feel better after you are dry?
4. Think about times you perspire. What do you think the perspiration accomplishes?

TEACHER INFORMATION

Let students discuss their responses to the above questions. Evaporation is a cooling process. When the body perspires, the evaporation of the moisture cools the skin, helping to control body temperature.

If some of the students have been around horses, they will know that when a horse runs it perspires. Some might also know, from experience or from their reading, that pigs are attracted to waterholes on hot days. Pigs do not perspire and this seems to be their way of getting the advantage of the cooling effect of evaporation. Many homes in dry climates are cooled by evaporation coolers, which operate on the same principle. Air is blown through water-soaked filters and the evaporative action cools the air, which in turn cools the home.

The skin helps control body temperature in other ways, too. Have students recall how flushed their faces become when they are hot. This occurs as blood vessels expand, allowing more of the heated blood to flow into the skin to be cooled. Then have them think about the "goose bumps" that form on their skin when they become cold. These occur as blood vessels contract, closing pores tightly to prevent body heat from escaping.

ACTIVITY 93: What Do Our Bones Do for Us?

MATERIALS NEEDED

- Model of a human skeleton

PROCEDURE

1. Examine the skeleton model.
2. Point to where the heart would be located if included in this model.
3. What bones do you see that surround the heart and protect it?
4. What other organs can you think of that are protected by bones? Point to them on your own body and point to where they would be if they were included in this model.
5. Notice how the backbone is constructed. Feel the separate bones in your own back. Why are there so many instead of just one long backbone?
6. Examine the arm and leg bones in the model. Notice how strong they are and where they bend. How do the joints help us?
7. Look at the hands and feet. How many of these bones can you find in your own hands and feet?

TEACHER INFORMATION

Although some students get an eerie feeling looking at a skeleton, the experience will help them to realize what they really are like beneath the skin. Some of the most obvious functions of the skeletal structure become evident as one examines a model of a human skeleton. Our bones support the flesh and give it shape. Joints are conveniently located to allow the body to bend. Many of the vital organs—heart, lungs and brain—are enclosed in protective coverings of bone.

ACTIVITY 94: Are Human Bones Large Or Small?

MATERIALS NEEDED

- Encyclopedia or other appropriate reference books
- Ruler
- Pencil and paper

PROCEDURE

1. Make a scale drawing of your femur (upper leg bone). Use a scale of 1 to 4 and check the encyclopedia for help with the shape of the bone.
2. Find the size of the same bone of various animals, such as dog, cat, rabbit, mouse, horse, and maybe one of the large dinosaurs. Use the encyclopedia as a source of information. Perhaps you can actually measure this bone of some animals if you have pets or farm animals.
3. Make a chart showing the comparative sizes of the femur—yours and those of the animals you used.
4. Is your bone large or small?

TEACHER INFORMATION

The student will find that size is relative and that human bones could be considered either large or small, depending on the size of the animals with which they are being compared. Students might prefer to make scale drawings of the bodies of various animals, instead of a single bone, and chart these in a way similar to that suggested above.

The lower grades could do this activity by substituting a general comparison of sizes for scale drawings. When students find information in a book about the size of a bone, have them locate something in the classroom or outside that is about the same size, to help them visualize it.

ACTIVITY 95: How Many Bones Can You Count?

MATERIALS NEEDED

- Paper and pencil
- Encyclopedia or other appropriate reference books

PROCEDURE

1. Feel the bones in the fingers of one hand with the other hand. Count them.
2. How many bones did you count in your hand? Write that number down.
3. See how many bones you can count from your fingers to your shoulder. Write that number down.
4. Now begin with your toes and work up as you count all the bones you can find. As you count the bones in your foot, leg, back, and so on, write down the numbers.
5. Using this procedure, count the bones in your entire body. As you write the numbers, remember to include the number of bones in both hands, both feet, and so forth.
6. Draw a picture of the human skeleton, including the bones you found.
7. If others in your group do this activity, compare your notes and drawing with theirs. If your numbers are different for some part of the body, each of you count again and try to determine where the difference occurred.
8. When you have counted and drawn all the bones in your body that you can find, go to the encyclopedia and see if you can find out how many bones there really are in the human body.
9. How close was your count?

TEACHER INFORMATION

The human body has at least 206 bones. A textbook that has this information could be substituted for the encyclopedia suggested in the materials list. The process of counting and searching the text or encyclopedia will provide a worthwhile and interesting research experience for students.

Lower grades can count, talk about, compare, and draw the bones they think they feel. The use of reference books can be eliminated for students who are not able to use them.

ACTIVITY 96: How Can Doctors Tell If and Where a Bone Is Broken?

MATERIALS NEEDED

- X-rays of broken and unbroken bones

PROCEDURE

1. Hold the X-ray films up to the window, one at a time, and compare.
2. Can you tell which bone is broken?
3. Do you see any cracked bones?
4. Why do you think the doctor puts a cast on arms and legs when bones are broken?.
5. What do you think might happen if an unqualified person moves someone who has been involved in an accident?

TEACHER INFORMATION

Before beginning ask if anyone in the class has ever broken a bone.

X-rays should be available at a hospital or doctors' clinic if you ask ahead of time and request that some be saved for use in class. Discuss the above questions. Ask students to consider question 5 in terms of a possible broken leg bone, arm bone, or rib bone. You might also wish to discuss it with respect to broken backbones, although this deals with very different and more technical issues.

ACTIVITY 97: How Is a Splint Applied to a Broken Bone?

MATERIALS NEEDED

- Newspapers
- Several strips of rags at least 5 cm. (2 in.) wide and 60 cm. (2 ft.) long

PROCEDURE

1. Pretend your leg bone is in one piece and it is not supposed to bend at the knee. Let the knee represent a break in the bone.
2. Have your partner use newspapers and rag strips to make a splint for your leg. Several sections of newspapers should be wrapped around the leg to make it stiff. A bone must not bend at a break, so be sure it is tied securely. It must not be so tight that it stops the flow of blood.
3. Try walking. Can you get yourself up? Does the leg feel that it would remain stiff so the bone could heal properly?
4. Trade places and make a splint for your partner's leg.
5. Have your partner stand up and test your splint to see if it feels secure.

TEACHER INFORMATION

In this activity students practice making a splint that will hold the bone securely without shutting off the blood supply. A first-aid manual would be an excellent resource to have for reference on splint making, but let students try to accomplish the task with minimal interference.

ACTIVITY 98: How Are Bones Connected at a Joint?

MATERIALS NEEDED

- Knee joint of an animal
- Encyclopedia
- Pencils

PROCEDURE

1. Examine the joint.
2. Can you see what holds the two bones together?
3. Use your pencil to probe around on the bone near the joint, in the joint, and away from the joint.
4. Is there any difference in the way the material feels with your probe as you move from one part to another? Where is it harder? Where is it softer?
5. Look up "bone" in the encyclopedia and find names for the parts you see, including what holds the two bones together at the joint.
6. Where do you think you have bones similar to those you are looking at?

TEACHER INFORMATION

Bones for this activity are usually available at meat markets. A front knee joint of a calf or sheep would work well, but most any joint will do. In checking the encyclopedia, students should be able to identify the bone, cartilage, and ligaments. Connecting points of tendons might also be visible.

As they probe with their pencils, students should be able to feel the softer cartilage material that cushions the bones at the joint.

Consider asking the butcher for a second joint, sawed lengthwise through the bone and joint, with which students can see where the cartilage is fused to the bone. Marrow will be in the center of the long part of the bones, providing an excellent research topic with the encyclopedia and other available sources.

ACTIVITY 99: What Are Tendons and How Do They Work?

MATERIALS NEEDED

- Chicken leg, complete with foot

PROCEDURE

1. Pick up the chicken leg and locate the tendons. The tendons are like cords and should be visible at the top of the leg.
2. Pull on the tendons one at a time and observe the foot.
3. What happened?
4. Try it again. What do you think the chicken does to curl its toes? What does it do to straighten them out?
5. How does this compare with the way your own fingers and toes work?
6. Locate some of the tendons that operate your fingers. See if you can tell where they connect and what makes them work. Wiggle your fingers quickly and watch the action of the tendons.
7. Grasp the large tendon at the back of one ankle with your fingers. Feel it as you move your foot up and down. This is called the Achilles tendon. Which muscles pull on the Achilles tendon?

TEACHER INFORMATION

Chicken legs and turkey legs work equally well for this activity. Be sure the tendons have not been removed. If the tendons are not visible at the top of the leg, cut the skin back to expose enough of the tendon for students to grasp. If some students are squeamish about operating the chicken foot, have others demonstrate for them.

Students should be able to easily identify similar structures on their own hands and feet and observe the tendons that attach muscles to bones. With careful observation they can tell which tendons open each finger and toe and about where they attach to the muscle. Point out that most muscles are attached to bones, either directly or by tendons.

Enrichment: Have a student, or group of students, do some research on the mythology of the Achilles tendon and report to the class.

ACTIVITY 100: What Is Bone Like Without the Animal Or Mineral Material?

(Teacher-supervised activity)

MATERIALS NEEDED

- Chicken leg bones
- Vinegar
- Metal pan
- Gram balance (or other sensitive scales)
- Heat source **(CAUTION: To be used only under teacher supervision)**

PROCEDURE

1. Soak one of the chicken bones in vinegar for four or five days.
2. Remove the bone from the liquid and dry it off.
3. Feel the bone. Bend it. What happened? What does it feel like?
4. Place the other bone on the metal pan and heat it until the bone is covered with a grayish-white ash. Weigh the pan and bone before you heat it.
5. Let the pan and bone cool.
6. Weigh the pan and bone and compare with the weight before heating.
7. Feel the bone. Bend it. What happened? What does it feel like?
8. Compare the heated bone with the one soaked in vinegar. How are they alike? How are they different?
9. From one of the bones you removed the animal material, leaving only mineral. From the other you removed the mineral. Which do you think is which?
10. How do you think the materials these bones are made of compare with your own?

TEACHER INFORMATION

A foil pie tin is adequate for heating the bone in step 4. The heating process burns off the animal material, leaving only brittle minerals. The bone will be much lighter after the animal material is removed. The vinegar, on the other hand, dissolves the minerals, leaving only animal material. This bone will be soft and flexible enough to tie in a knot.

ACTIVITY 101: How Many Muscles Can You Identify?

MATERIALS NEEDED

- Paper
- Pencils

PROCEDURE

1. Raise your arm slowly. As it moves, try to identify the muscles that make it move.
2. Write down the movement and describe where you think the muscles causing the movement are located. For instance:

Movement	Muscle Location
Raise the arm	From shoulder to top of upper arm

3. Lower the arm onto a table. Then push against the table top in an effort to lower the arm further. Find the muscles that seem to pull the arm down.
4. Write down the movement and describe where you think the muscles causing the movement are located.
5. Continue this for all the different arm movements you can think of. Do the same with the hand, then the legs and feet, then other body parts.
6. Compare your list of body movements and muscle locations with others. See how many more you can identify altogether.

TEACHER INFORMATION

Essentially every movement of the body is produced by muscular action. Since muscles pull (contract) but do not push, a different set of muscles is used for opening the fingers than for closing them. The same can be said for many other body movements, such as moving the leg forward and backward, raising and lowering the arm, and so on. (Gravity should be taken into account.) Where ball joints are involved, muscular arrangements also allow a twisting motion.

The human body has over 650 different muscles. Students will be able to locate many of these as they examine their own body movements. There is some advantage in putting students in groups of two or three for this activity so they can analyze their movements together, discuss their observations, and learn from each other.

ACTIVITY 102: How Do Voluntary and Involuntary Muscles Differ?

MATERIALS NEEDED

- Mirror

PROCEDURE

1 Do the following:
   a. Close one hand and open it.
   b. Lift one foot and put it down.
2. Did these muscles move because you decided to move them? Do they ever move other than when you decide to move them?
3. Do the following:
   a. Look in the mirror and watch your eyes. Notice the size of the pupils (black spot in the middle). Shade one eye with your hand as you observe the pupil.
   b. Put your hand over your heart and feel it beat
4. Did the size of the pupil change? Did you decide to change it? Can you change the size of the pupil without changing the light? Can you make your heart beat just when you want it to or make it beat faster or slower?
5. How does your control over the muscles you used in step 1 compare with your control over the muscles in step 3?
6. Now do the following:
   a. Look at your eyes in the mirror for 30 seconds. Did they blink? Do they blink even if you don't decide to make them blink? Can you make them blink faster or more slowly?
   b. Notice how fast you are breathing. Does this happen even if you don't think about it? Can you breathe faster or more slowly if you want to?
7. Compare what you did in steps 1, 3, and 6. How do they compare in the amount of control you have?

TEACHER INFORMATION

Students will discover some body movements are controlled by voluntary muscles (step 1) and some by involuntary muscles (step 3). Still other muscles are both voluntary and involuntary (step 6). For instance, we can speed up our breathing or the blinking of the eyes, but if we don't think about it, automatic mechanisms take over. Breathing and blinking go on without conscious effort on our part. We can also delay these actions temporarily, but if we interfere too long, the involuntary actions will override our efforts.

Many involuntary muscles are constantly at work inside our bodies. The stomach and intestines contract and relax to aid digestion and to move food material along the digestive tract. Arteries contract and relax to help move the blood to various parts of the body. These processes take place regardless of any conscious effort on our part.

Not all people have equal control of voluntary muscles. Differences are evident when considering handicaps, such as palsy, which have a wide range of effects on muscular control. It should also be pointed out that even among those not considered handicapped, there are great differences in the ability to control the muscles. This is, in part, responsible for varying abilities in art, athletics, and many other skills.

ACTIVITY 103: What Is Muscle Sense?

MATERIALS NEEDED

- Blindfold

PROCEDURE

1. Blindfold your partner.
2. Place your partner's left arm in a raised position and ask him or her to hold it there. Then instruct your partner to put the right arm in the same position as the left.
3. Was your partner able to match the position of the left arm with that of the right arm?
4. Move your partner's left arm to a different position and again ask him or her to put the right arm in the same position.
5. Repeat this process several times, sometimes positioning the right arm and asking your partner to match its position with the left arm.
6. Is your partner able to match the position of one arm with the other each time without looking? Why do you think this is so?
7. Trade places. You wear the blindfold and have your partner test your ability to match the position of one arm with the other.

TEACHER INFORMATION

Certain nerves leading from the muscles to the brain tell the position of the muscles. This is called *muscle sense.* As a result of muscle sense, people have automatic knowledge of the position of the muscles.

Section 7

THE FIVE SENSES

TO THE TEACHER

The human body is a topic of interest, curiosity, and importance to all ages. Formal study of it should begin in the elementary grades. Many things can be done at this early age to increase awareness of the capacities and needs of this marvelous system. As awareness increases, so do appreciation and the ability to care for our bodies properly.

Everything we do involves one or more of the five senses. All that we learn is learned through the senses. Getting acquainted with their bodies is a logical topic for young learners. Scores of activities can be undertaken that involve concrete, first-hand experiences. Many concepts have been encountered before, but new insights and awarenesses should be acquired as those concepts are spotlighted and discussed.

A study of the five senses should include recognition of the handicapped. Those who have lost part, or all, of one or more senses deserve to be recognized and respected as normal human beings. Children should develop an attitude of appreciation for their capabilities without perceiving the handicapped as something less. Indeed, people with full capability of the senses can learn a great deal from those with some degree of loss of hearing, sight, or other capabilities. Frequently, other senses have compensated by becoming sharper, resulting in enhanced awareness.

ACTIVITY 104: How Does a Picture Stimulate Your Senses?

MATERIALS NEEDED

- Pictures from books and magazines

PROCEDURE

1. Find a picture in a book or a magazine.
2. Tell some ways that what you see involves the senses of touch,hearing, smell, or taste.
3. Share your ideas with others and make a list.

TEACHER INFORMATION

You might wish to use this as an introductory activity or as a culminating activity for a study of the senses. A winter scene plays on the sense of touch as one infers from it cold temperatures and scratchy branches on the trees. A picture of delectable food stimulates the senses of taste and smell. When we look at a scene of a busy street corner or a football game we can almost hear the honking horns or the roar of the crowd as a touchdown is made by their team. Artists and photographers are able to enhance perceptions and enjoyment by involving more than the sense of sight on the part of the observer.

Students might also enjoy drawing their own pictures to see how many of the senses they can involve.

Discussion accompanying this activity should include not only the inference of other senses from the sense of sight, but the intertwining of all the senses and their ability to stimulate each other. For instance, when we hear something sizzling on a stove before breakfast, we might visualize bacon and eggs frying.

Publishers have added the sense of smell to pictures in a very real way with "scratch and smell" pictures. Consider adding some of these to the discussion of this activity.

ACTIVITY 105: What Foods Can Your Nose Identify?

MATERIALS NEEDED

- Variety of food samples
- Paper cup for each sample
- Toothpicks
- Blindfold
- Markers
- Chart paper

PROCEDURE

1. Choose a partner to help you with this activity.
2. Blindfold your partner.
3. Select one of the food samples and use a toothpick to hold it under your partner's nose for a few seconds. Ask him or her to identify the food.
4. Record your partner's response and indicate whether the food sample was identified accurately.
5. Follow the same procedure for the remaining food samples. For any food sample your partner did not identify correctly, use a toothpick and place a small amount on your partner's tongue. Then see if he or she can tell what the food is.
6. Trade places and ask your partner to give you the same food identification test.
7. Examine the charted results and compare your chart with your partner's chart. Which foods did his or her nose identify correctly without help? Which did yours? Were there any differences? If so, what ideas do you have about the reasons?

TEACHER INFORMATION

This activity will be more revealing if food samples with varying strength of odors are selected. Students will find that some people can smell certain odors better than others, and that the nose relies, at times, on assistance from the tongue, just as the sense of taste is sometimes assisted by the sense of smell.

ACTIVITY 106: How Fast Do Odors Travel?

MATERIALS NEEDED

- Bowl or saucer
- Perfume or after-shave lotion
- Timer or clock with second hand

PROCEDURE

1. Have the students on one side of the room put their heads down and close their eyes.
2. At the front of the room, put a few drops of the perfume in the bowl and ask the students with their heads down to raise their hands when they smell the perfume. They are not to open their eyes until told to do so.
3. Record the number of seconds it takes for the aroma to reach the first row, second row, third row, and so on.
4. When you are finished, have the students open their eyes.
5. Discuss with the class the results of the investigation.

TEACHER INFORMATION

This activity will show not only the speed with which aromas move through the air, but the differences in sensitivity of the sense of smell from one person to another. Students might wish to test and compare different perfume brands and fragrances. If so, the bowl will need to be rinsed and the room aired out between trials.

Use the other half of the class to repeat the activity, adding plain water as a secret brand of perfume. Find out how many students "smell" perfume just because they think it's there.

ACTIVITY 107: How Long Can You Retain a Smell?

MATERIALS NEEDED

- Slice of orange
- Blindfold

PROCEDURE

1. Find a partner to do this activity with you.
2. Hold the slice of orange under your partner's nose.
3. Tell your partner to close his or her eyes and to tell when you move the orange slice away.
4. Leave the orange slice under your partner's nose for two minutes, or until he or she reports that it has been removed.
5. What happened? Try to explain why.
6. Try the same activity with other people and with substances other than the orange slice.

TEACHER INFORMATION

Other items can be substituted for the orange slice, but they should have a distinct odor and the odor should not be too strong. The odor-sensitive nerves seem to become accustomed to a given smell in a short time and cease to recognize that odor. The person smelling the orange slice or other substance will usually report in a short time that the item has been removed even though it is still there.

This is a good time to develop or practice graphing skills. Have students make a graph showing smell duration of several people, using the same substance. Other graphs could be made using a variety of substances with the same person.

ACTIVITY 108: What Differences in Taste Do People Have?

MATERIALS NEEDED

- Variety of food samples
- Paper cups (one for each type of food)
- Box of toothpicks
- Blindfold
- Markers
- Chart paper

PROCEDURE

1. Find a partner to help you with this activity.
2. Blindfold your partner.
3. Using a toothpick, place a small amount of one type of food in the center of your partner's tongue.
4. Have your partner close his or her mouth and move the tongue around for a thorough taste of the sample. Then describe the food as sweet, sour, salty, or bitter. Record the results on a chart like the one shown:

   Name of taster: ____________________

FOOD	TASTE			
____________	Sweet	Sour	Salty	Bitter

5. Follow the same procedure for each of the food samples, using a new toothpick for each food. The person tasting should rinse his or her mouth with water between food types. Place the food in the center of the tongue each time.
6. After your partner has tasted all the food samples, trade places and have your partner put the food samples on your tongue. Put the food in the center of the tongue each time and use a new toothpick for each type of food. Your partner should record your judgment of each food type as sweet, sour, salty, or bitter.

TEACHER INFORMATION

Tastes are usually described as sweet, sour, salty, or bitter. These tastes are strong in some foods and weak in others. Some foods even have combinations of tastes. The purpose of this activity is to compare taste judgments of different people. The different taste-sensitive areas of the tongue are compared in Activity 110, and should not become a factor in the above activity; for consistency it is important that each food sample be placed at the center of the tongue. You might want to make several copies of the record chart for students to use.

ACTIVITY 109: How Is Taste Affected by Smell?

MATERIALS NEEDED

- Same food samples as those used in Activity 108
- Paper cups (one for each type of food)
- Box of toothpicks
- Blindfold
- Record charts from Activity 108
- Markers

PROCEDURE

1. Find a partner to do this activity with you.
2. Blindfold your partner.
3. Have your partner hold his or her nose so the foods being tasted cannot be smelled.
4. Follow steps 3 through 6 from Activity 108. Record taste judgments beside the judgment made by the same person for each food type in Activity 108. Use a different-colored marker from that used in Activity 108 so the two sets of information will not get confused.
5. When finished, compare each person's taste judgments with those of the same person in Activity 105. Are there any differences? If so, what do you think made the difference?

TEACHER INFORMATION

People are sometimes heard to say that food just doesn't taste the same when they have a cold. Taste is affected by smell, sometimes enough to alter the judgment of the type of taste if the salty, sweet, sour, or bitter tastes are not very strong. Have students share their findings and discuss them.

ACTIVITY 110: Which Part of the Tongue Is Most Sensitive to Taste?

MATERIALS NEEDED

- Variety of food samples
- Paper cups (one for each type of food)
- Box of toothpicks
- Blindfold
- Water
- Chart paper
- Markers

PROCEDURE

1. Find a partner to do this activity with you.
2. Blindfold your partner.
3. Using a toothpick, place a small amount of one type of food on the region of the tongue identified as "1" in the illustration. Your partner is to judge the taste with his or her mouth still open so the food sample is not spread to other regions of the tongue. The taste judgment this time is to indicate strength as well as type of taste: strong sweet, weak sweet, strong salty, weak salty, strong sour, weak sour, strong bitter, or weak bitter.
4. Record your partner's judgment of taste, have him or her rinse mouth with water, then place the same type of food on region 2, then 3, then 4, then 5. Record the taste judgment each time. Be sure the mouth is rinsed with water between tastes.
5. When you have placed the first food type on all five regions of the tongue and recorded your partner's judgment of taste, do the same with the next food type.
6. After recording your partner's taste judgment of each type of food, trade places and have your partner give you the same taste tests.
7. Analyze the information you have collected in this activity and see if some regions of the tongue seem to be more sensitive to certain tastes than other regions. Make a chart showing your findings.

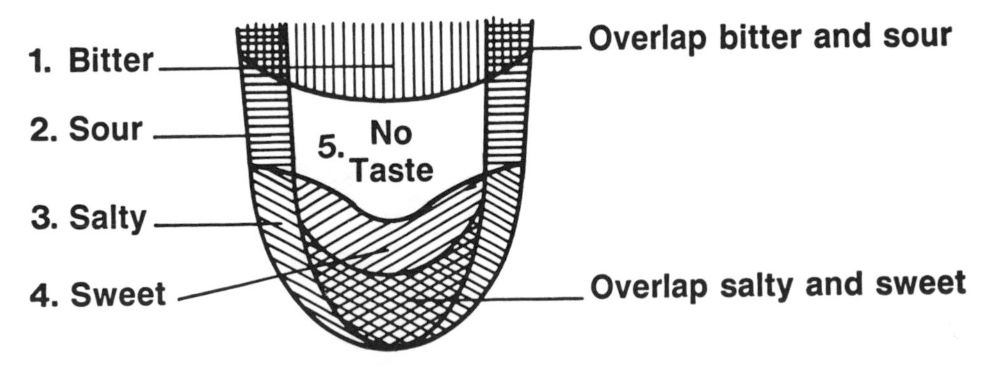

FIGURE 110-1. Taste regions of the tongue.

TEACHER INFORMATION

Certain regions of the tongue are known to be more sensitive to certain tastes. These are shown in Figure 110-1. Results of the above investigations should approximate this information. Have students share and compare their findings. Where differences are found, encourage students to try to identify reasons for the differences. Possible factors include not placing the food sample in the exact same locations, closing the mouth, or spreading the food sample to other parts of the tongue. Also consider differences in taste sensitivities from one person to another (refer to Activity 108), or simply the subjectivity of the judgment.

ACTIVITY 111: How Can You Tell If Taste Is Affected by Smell?

MATERIALS NEEDED

- Small slices of potato, apple, and onion
- Three blindfolds
- Chart paper
- Markers

PROCEDURE

1. Find three volunteers to help you do this activity.
2. Blindfold all three volunteers. Seat them about two meters (six feet) apart.
3. Hold a slice of one of the three foods under the nose of one volunteer as you put another of the foods in the same person's mouth. Ask him or her to identify the food in the mouth.
4. Use various combinations of the food slices with the same volunteer, placing a slice of apple in the mouth as you hold a slice of onion under the nose. Place a slice of potato in the mouth as you hold a slice of apple under the nose, and so forth.
5. Record the foods used and what the volunteer identifies them to be.
6. Follow steps 3 to 5 with the other two volunteers.
7. Analyze and discuss the results. Did the volunteers identify the foods accurately in all cases? Were the results consistent from one volunteer to the next? If identifications were inaccurate in some cases, which smells seem to fool the tongue?

TEACHER INFORMATION

The senses of taste and smell work together in helping us to identify what we are eating and to enjoy it. Either of these senses can be affected by strong signals from the other. Students might enjoy trying this same activity with other combinations of foods.

ACTIVITY 112: How Can We Classify Foods by Taste?

MATERIALS NEEDED

- Variety of foods
- Chart paper
- Markers

PROCEDURE

1. For a period of one day, keep a list of all the food items you eat.
2. As you eat something, think about ways you could describe its taste.
3. As you write each item in your list, write a description of its taste. A word or two is usually sufficient.
4. When you have finished your list, look it over and find foods that seem to have similar taste. Group these together.
5. Compare your list with those of others. Does someone else have some of the same foods? Did they describe the taste the same way you did? Talk together about it.

TEACHER INFORMATION

If students cannot think of ways to describe taste in order to classify the foods they eat, suggest or discuss the terms normally used to describe taste: sweet, sour, salty, and bitter.

A fun way to do the above activity is to have a party to which each member of the group contributes one food item. Then before anything is eaten have students list all food items involved and classify each one as either sweet, sour, salty, or bitter. They could work individually, in small groups, or as a whole class.

ACTIVITY 113: What Sounds Do You Hear in Paper?

MATERIALS NEEDED

- Sheets of paper

PROCEDURE

1. Have all your students put their heads down and close their eyes.
2. Make sounds with the paper and have the group try to guess what you are doing with the paper to make each sound. For instance, fold it, cut it, tear it, crumple it up, shake it, drop it on the floor, smooth out the crumpled paper, blow on it.

TEACHER INFORMATION

This activity lets students use their sense of hearing and their experience with the sounds that can be made with paper. Students might want to take turns making one sound at a time to see if the others can determine what is happening to the paper using the sense of hearing alone. Have them try different kinds of paper, such as ditto paper and construction paper. See if the group can tell what kind of paper is being used as well as what is being done with it.

ACTIVITY 114: What Sounds Do Your Ears Recognize?

MATERIALS NEEDED

- Paper
- Pencils

PROCEDURE

1. Close your eyes and listen for about three minutes.
2. Write a list of all the sounds you heard and their sources.
3. Discuss this experience with others. Did you hear the same things everyone else did? Did you notice sounds that you don't usually notice? Why did some sounds register this time when they sometimes go unnoticed?

TEACHER INFORMATION

This should be a group activity, as the sharing of sounds heard and ideas about them is an important part of the experience. Students will note that when they listen deliberately for all sounds they will be aware of sounds that frequently go unheeded.

As a follow-up to this activity, consider using a tape recorder to capture a variety of sounds, such as a blender running, the shutting of a refrigerator door, ice cubes being broken from their tray, a busy street corner, or other common sounds. Play the tape to the class and see how many of the sound sources they can identify.

ACTIVITY 115: How Much Can You "See" with Your Ears?

MATERIALS NEEDED

- Curtain or other opaque barrier
- A "sound" skit
- Props (optional)

PROCEDURE

1. Find two or three partners to do this activity with you.
2. Create a short skit, such as a trip to the grocery store, going swimming, or mountain climbing. Your skit is to be heard only, not seen, so write the speaking parts and include any sound effects that might help.
3. Behind the curtain, perform your skit for your class.
4. Tell the class that after your skit is finished you'd like them to describe the events of the play. The class members are to use only their sense of hearing, as all actions are hidden from view.
5. Perhaps other groups in the class would like to try producing a "sound skit."

TEACHER INFORMATION

Usually, skits are both seen and heard, so we use two senses in getting the meaning of the play. In this activity, students are to use only their sense of hearing. Encourage them to close their eyes as they "observe" with their ears, and try to visualize the scenes being acted out in sound.

ACTIVITY 116: How Well Do You Know Your Classmates' Voices?

MATERIALS NEEDED

- Blindfold
- Chair

PROCEDURE

1. Do this activity with a group of classmates.
2. Place a chair at the front of the room.
3. Select a volunteer to sit on the chair. This is person 1.
4. Blindfold person 1.
5. Choose a person (person 2) to come up and stand behind person 1.
6. Person 2 knocks on the back of the chair person 1 is sitting in.
7. Person 1 asks, "Who is knocking?"
8. Person 2 answers, "It is I," with a disguised voice.
9. Person 1 has three chances to guess who answered, "It is I."
10. If the guesses are all wrong, another person is selected to be person 2. If the guess is right, person 2 becomes person 1, sits on the chair, and is blindfolded. Another person is selected to come up and knock on the chair.

TEACHER INFORMATION

This is an exercise in using the sense of hearing, coupled with familiarity with the voices of friends in the class. The student in the role of person 2 needs to come up to the front of the room very quietly so person 1 doesn't know what part of the room he or she came from. The challenge is trying to determine whose voice is being disguised.

ACTIVITY 117: How Can You Match Things without Seeing Them?

MATERIALS NEEDED

- Bean seeds (at least 10)
- Small erasers (at least 6)
- Paper clips (at least 10)
- Thumb tacks (at least 10)
- Buttons (at least 10)
- 10 small cans with lids

PROCEDURE

1. Place half the bean seeds in each of two cans and put the lids on.
2. Do the same with the erasers, paper clips, thumb tacks, and buttons.
3. Be sure all lids are on tight. You should now have two containers with bean seeds inside, two with erasers, two with paper clips, two with buttons, and two with thumb tacks.
4. Mix the cans up.
5. Shake the cans one at a time and try to match them in pairs, according to what they have inside. Do not look inside the lid to match contents. Use only your ears to pair them up.
6. Have someone who doesn't already know what is inside try matching the shaker cans.

TEACHER INFORMATION

We frequently classify things according to visual characteristics. This activity requires the use of the sense of hearing only in matching pairs of cans containing identical items. Other small items can be substituted for those listed above, and the cans can be replaced by other suitable containers if necessary or desirable.

As an interesting variation, use an odd number of cans and have students identify the can that does not have a mate. You might also try having only one matching pair in the entire set, with the objective being to identify the matching pair. The activity can be further extended by having students try to guess what is in each container. If students already know what the items inside the containers are, replace one or more of them with something else.

If you are using containers with transparent or translucent lids, glue or tape a paper inside the lid so the contents cannot be seen. Also consider decorating the outside of the containers. This could be done as an art activity before the items are put inside.

ACTIVITY 118: How Well Can Your Ears Alone Tell You What's Happening?

MATERIALS NEEDED

- Paper and pencils
- Carrot and grater
- Chalkboard and chalk
- Other optional props

PROCEDURE

1. Do this activity with several partners.
2. Ask the other participants to close their eyes and put their heads down.
3. Instruct them that they are to listen and try to decide what you are doing by using only their ears. They may raise their hands when they think they know.
4. Walk across the floor and see who guesses first what you are doing.
5. Write on the chalkboard.
6. Grate a carrot.
7. Open a drawer.
8. Do a variety of other things and see if the others can tell what you are doing.
9. Make a list of the things the others could identify by using only their ears and a list of things you did that they could not.
10. Examine your two lists and discuss reasons you think some of your actions were easier for the others to guess and some were more difficult.

TEACHER INFORMATION

Because of our familiarity with some sounds, certain ones are easily identified. Some are less obvious to us because we don't hear them very much or because they are similar to other sounds we know. This activity should emphasize to students that, although we often need the help of our eyes and other senses, our ears alone frequently provide us with rather accurate information.

You might want to discuss with the class how certain senses are sharpened when others are lost or weakened. The blind, for instance, learn to rely more on their ears, and they notice sounds that the rest of us sometimes do not.

ACTIVITY 119: How Effective Is Your Side Vision?

MATERIALS NEEDED

- Pencil or ruler

PROCEDURE

1. Do this activity with a partner.
2. Have your partner sit down and look straight ahead.
3. Hold the pencil vertically about one meter (40 inches) away from your partner's ear.
4. Move the pencil forward slowly and ask your partner to tell you when he or she can see it (cyes still straight ahead).
5. Move the pencil slowly forward and back as necessary to find the point at which your partner can first see it.
6. Test your partner's side vision of the other eye.
7. Trade places and have your partner help you test your side vision.
8. Compare results. Are they the same? Discuss situations where good side vision might be important.
9. Have you ever had an eye specialist perform this examination?

TEACHER INFORMATION

Peripheral vision (side vision) is the ability to see at the side while focusing straight ahead. It is used frequently by everyone, but for some it is narrower than for others. In some situations, good peripheral vision is critical, such as when driving an automobile or walking across a busy street. Students should easily recognize the importance of side vision to a football player or basketball player. Many other examples can be discussed.

ACTIVITY 120: How Well Do You Remember what You See?

MATERIALS NEEDED

- Tray
- Variety of small objects

PROCEDURE

1. Put a variety of small objects on the tray, such as a pencil, eraser, marble, paper clip, toy car, bracelet, wad of paper.
2. Ask one or more participants to examine the items on the tray for 30 seconds.
3. Remove the tray from sight. Remove one of the items.
4. Return the tray and ask participant(s) to look over the contents and try to determine which item was removed. Ask them not to say it aloud until all participants have decided what they think it is.
5. Did everyone get it right? Did no one get it right?
6. If time allows, do the same activity again with the same group. See if they can, with practice, improve their visual memory skills.

TEACHER INFORMATION

This activity is effective in evaluating visual memory skills. With repeated use, it also provides practice in using visual memory. As skills improve, try increasing the number of objects or rearranging them.

ACTIVITY 121: In What Ways Do We Depend on Our Eyes?

MATERIALS NEEDED

- Pencils
- Two sheets of paper for each student
- Blindfold

PROCEDURE

1. On one of the papers, draw a simple picture and write your name at the bottom of the paper.
2. Have someone blindfold you. Then using the other paper, draw the same picture again and write your name on it in the same place you did on the first.
3. Compare your pictures. How well were you able to draw the picture when you couldn't see?
4. Name some other ways in which we depend on our eyes.

TEACHER INFORMATION

As students compare their pictures and consider the difficulty in placing elements of the picture in the right places and connecting the lines, their dependence on their eyesight should be emphasized. They will probably think of many ways and times we depend heavily on our eyes.

ACTIVITY 122: How Do Our Eyes Help Us "Hear"?

MATERIALS NEEDED

- Storybook with pictures

PROCEDURE

1. Get a group of your fellow students to help you with this activity. They will be your listeners.
2. The storybook you choose should have interesting pictures and illustrations. It should be one the listeners have not heard or read before.
3. Have half of the listeners close their eyes and keep their heads down while the story is read to the group.
4. Read the story to the group, showing the pictures to those who have their eyes open. Don't talk about the pictures.
5. Discuss the story. Let those who had their eyes closed tell about it first—then see if those who had their eyes open can add any information or details the others were not aware of.
6. If those who had their eyes open could add information the others did not know, discuss the reason. If not, discuss the author's ability to help listeners to form word pictures in their minds.

TEACHER INFORMATION

The story in this activity could be read to the class or to a small group of students by the teacher or by one of the students.

Discuss ways our ears are often assisted by our eyes—not only as we read and listen to stories but also at the many other times our eyes and ears help each other.

Students might wish to find a second story and change roles, having the other half of the group close their eyes.

ACTIVITY 123: What Happens to the Iris As Light Changes?

MATERIALS NEEDED

- Mirror

PROCEDURE

1. Hold the mirror close enough to your face so that you can easily see the iris (colored part) of your eyes.
2. Look at the iris of one eye carefully and see if you can detect any movement.
3. Hold one hand up to the side of your eye to shade the light. As you do, watch the iris carefully.
4. Remove your hand from the eye, still watching the iris.
5. Close one eye or put your hand over it to shut out the light. When you open it, observe the iris immediately.
6. What happens to the iris as you change the amount of light around it? Why do you think this happens?

TEACHER INFORMATION

The iris opens and closes to adjust the amount of light entering the eye through the pupil. The diaphragm of a camera operates much the same way (see Activity 124).

You might pair students up and have them observe the iris in each other's eyes as light conditions change. Caution them not to shine bright light in their eyes or look directly at bright lights. The movement of the iris is easily observed in room light by closing the eyes or temporarily shading with the hand.

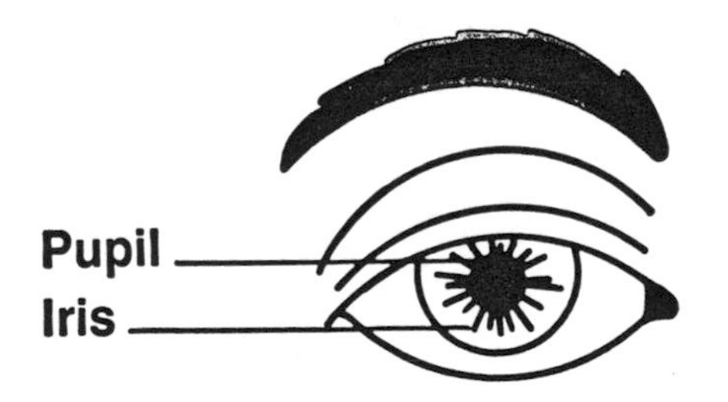

FIGURE 123-1. The iris of the eye.

ACTIVITY 124: How Is a Camera Diaphragm Like the Eye?

(Teacher-supervised activity)

MATERIALS NEEDED

- Camera

PROCEDURE

1. The camera must not have film in it and must have an adjustable diaphragm (f-stop).
2. Ask your teacher to be close by for this activity, to assure that no damage will be done to the camera and to help you locate the parts you need to work with.
3. Open the back of the camera.
4. Set the shutter speed on time (B) so it will remain open when the shutter release button is depressed.
5. Look through the lens from the back of the camera. You should be able to see a dot of light.
6. Move the f-stop setting as you look through the lens.
7. What happened? How does it compare with the movement of the iris of your eye you observed in Activity 123?

TEACHER INFORMATION

This activity should be carefully supervised. If you have to borrow a camera and are not acquainted with the operation of it, it will take only a few minutes for someone to brief you on it for this activity This is also an excellent time for inviting a resource person into your classroom if you have a camera buff available. They could talk about the light adjustments of the camera and compare these to the adjustments made by the eye. Some cameras use solar light cells to adjust the diaphragm automatically.

ACTIVITY 125: How Well Can You Judge Depth with One Eye?

MATERIALS NEEDED

- Ping pong ball
- Soda bottle
- Table (lower than waist high)

PROCEDURE

1. Stand the soda bottle on the table, about 15 cm. (6 in.) from the edge.
2. Place the ping pong ball in the top of the bottle.
3. Walk away at least 3 meters (10 ft.).
4. Face the bottle, cover one eye with your left hand, and walk toward the bottle.
5. As you pass the bottle, try to flip the ball with your finger. Flip only one time and do not pause to flip.
6. What happened?
7. Try it again, covering the other eye.
8. Were the results any different?
9. Try it a third time, leaving both eyes uncovered.
10. What happened this time? Compare the three attempts and explain as best you can.
11. Have someone else try flipping the ball, using the same procedures. Compare the results with your own.

TEACHER INFORMATION

In this activity, students will learn that having two eyes serves more of a purpose than simply providing a spare. Accurate depth and distance perception requires two eyes. It is usually difficult to flip the ping pong ball with one eye covered. Students might enjoy practicing to see if they can increase their skill. You might also have them chart the results of several attempts and find out if their accuracy is any greater with one eye than with the other.

ACTIVITY 126: Which Is Your Dominant Eye?

MATERIALS NEEDED

- None

PROCEDURE

1. Look at an object that is at least 3 meters (10 ft.) away from you.
2. With both eyes open, point at the object.
3. Without moving your pointing finger, close your left eye. Does your finger still appear to be pointing at the object?
4. Now open your left eye and close your right eye. Does your finger appear to be pointing at the object?
5. Select another object and repeat steps 2 through 4.
6. What happened? Try to explain why.

TEACHER INFORMATION

The two eyes, being a short distance apart, see objects from a slightly different angle. This has certain advantages, including helping us to perceive depth and distance with much greater accuracy than would otherwise be possible. When we point at an object, the finger is in the line of vision of only one eye. This is called the dominant eye, as it is nearly always the same eye. People with two good eyes usually have one that is dominant.

ACTIVITY 127: How Much Can You "See" with Your Hands?

MATERIALS NEEDED

- Paper
- Pencils
- Assortment of buttons (many shapes, sizes, and colors)
- Blindfold

PROCEDURE

1. Find a partner to do this activity with you.
2. Put the blindfold on your partner.
3. Pour the buttons out on the table and ask your blindfolded friend to separate them into groups according to whatever characteristic he or she chooses (such as size). Do not answer any questions about the appearance of the buttons. These characteristics must be determined by the sense of feel.
4. When your partner is finished sorting the buttons, ask him or her to try to think of another way to sort them (such as according to shape).
5. If your partner can think of still another way to sort the buttons, have him or her do it.
6. Write a list of the characteristics your partner uses for sorting the buttons.
7. Remove the blindfold and ask your partner to try to sort the buttons in still more ways. Add these to your list of characteristics. These might include color and texture.
8. After the blindfold was removed, what characteristics were used that were not used before? Could any of these have been felt, even though they weren't thought of until they were also seen? (How about texture, for instance?)
9. Can you think of other times when our sense of touch is aided by our sense of sight?

TEACHER INFORMATION

As a supplement to this activity, you might bring some things to class that feel similar but are not the same and do not look the same. Have blindfolded students try to decide what these things are. For example, try small jellybeans and chocolate-covered peanuts.

ACTIVITY 128: How Well Can You "Observe" with Your Sense of Touch?

MATERIALS NEEDED

- Paper
- Pencils
- Blindfold
- Variety of objects, such as a piece of sandpaper, cotton ball, orange, baseball, seashell, rock, feather, book, eraser, pencil, or other objects from around the classroom or brought from home

PROCEDURE

1. Find a partner to do this activity with you.
2. Blindfold your partner.
3. Hand one of the objects to your blindfolded friend and ask for a description of the object. The other person is not to name the object, but only to describe how it feels.
4. Write the name of the object and some of the descriptive words used by the blindfolded person.
5. Do the same with several other objects.
6. Have someone else try some of the same objects blindfolded and compare descriptions.

TEACHER INFORMATION

Descriptions given in this activity are to be limited to those characteristics of the objects that can be felt, such as smoothness, sharpness, roughness, size, shape, etc. If this is used as a whole-class activity, any items selected ahead of time should be put out of sight so each person, in turn, is experiencing some element of surprise. An alternative would be to select items for each person after the blindfold is in place.

Another option is to use a "feely box." Prepare a box that is closed except for an opening large enough for inserting the hand. A variety of small items is placed inside and students reach in, take one object in their hand (but keeping it inside the box), and describe what the object feels like. With this version of the activity the blindfold is not necessary.

ACTIVITY 129: Where Is Your Sense of Touch Most Sensitive?

MATERIALS NEEDED

- None

PROCEDURE

1. Do this activity with a partner.
2. Ask your partner to close his or her eyes.
3. Place the tips of three of your fingers, about 3 cm. (1 in.) apart on your partner's back. Ask your partner to tell you how many fingers are touching your back.
4. Touch the palm of your partner's hand with one finger, then two. See if your partner can tell without looking how many fingers you are using.
5. Do the same on the shoulder, the forearm, the neck, the forehead, and the back.
6. In what areas is your partner able to tell with the most accuracy how many fingers you are using?
7. Trade places and have your partner test your sense of touch in the same manner.
8. Compare the results. Is your sense of touch the most sensitive in the same areas as that of your partner?

TEACHER INFORMATION

Sometimes it is surprising to find out how little sensitivity we have on some parts of the body. Students usually find that on their hands, fingers, and face they can tell quite accurately how many fingers are touching. On the back and arms the sense of touch is noticeably less sensitive. They might like to discuss times when they have noticed that injuries are much more painful in some places than in others.

ACTIVITY 130: Why Do We Need Five Senses?

MATERIALS NEEDED

- Five blindfolds
- Five sheets of paper
- Five pencils
- Chart paper and marker
- Five baby food jars, each containing one of the following: salt, sand, granulated sugar, powdered sugar, and cornstarch

PROCEDURE

1. Choose five volunteers. Be sure they have not seen the jars containing the five substances.
2. Seat the volunteers at a table and blindfold them.
3. Place one of the jars in front of each volunteer. Also give each one a paper and pencil.
4. Ask each person to feel the contents of the jar and write on the paper what he or she thinks it is. The volunteers are not to taste it and they are not to say aloud what they think it is.
5. Record the written responses on a chart.
6. Rotate the jars one position to the right.
7. Again have the volunteers feel the contents and write down what they think the substance is. Record the results on the chart.
8. Continue until each of the volunteers has identified all five substances using only the sense of touch.
9. Be sure the chart is where it will not be seen by the volunteers and remove the blindfolds.
10. Place the jars in front of the volunteers in a different order from that of step 4.
11. Ask each volunteer to look at the substance in the jar and write what he or she thinks it is. The volunteers are not to taste or feel the substance. They are not to give their answers aloud, and they must not look at each other's responses.
12. Again rotate the jars, recording the responses of each participant.
13. When all five substances have been identified by all five participants by both touch and sight, let them use other ways to identify the substances.If they suggest tasting, assure them that none of these substances is harmful to taste.
14. Discuss the results. How accurate were the responses from the sense of touch alone? From the sense of sight alone? From a combination of these, and possibly with help from the sense of taste? How do the senses depend on each other? How do all five senses help us to know what is happening around us?

TEACHER INFORMATION

This activity should emphasize that the senses are interdependent. Discuss the fact that everything we learn is learned through the use of the five senses—frequently a combination of two or more of them. Also discuss how we rely on what we have already learned—information stored in the brain. We acquire certain information about these substances, for instance, by looking, touching, and tasting, but it is only from previous experience that we decide whether it is sugar, salt, or sand.

ACTIVITY 131: What Is Warm and What Is Cold?

MATERIALS NEEDED

- Three pans or bowls
- Hot, lukewarm, and cold water
- Blindfold

PROCEDURE

1. Find another student to do this activity with you.
2. Fill one pan with cold water, one with lukewarm water, and one with hot water. The hot water must not be hot enough to burn or hurt.
3. Blindfold your partner.
4. Have your partner put the left hand in the hot water and then put the right hand in the lukewarm water.
5. Ask your partner to describe the temperature of the water in the two bowls.
6. Have your partner remove both hands from the water.
7. Shift the bowls around, then have your partner put the left hand in the cold water and the right hand in the lukewarm water.
8. Ask your partner to again describe the temperature of the water in the two bowls.
9. Think about the result. Did your partner describe the lukewarm water as lukewarm both times? Did his or her judgment of the lukewarm bowl seem to be affected by the temperature of the water the other hand was in? How was it affected? What can you say about this?
10. Try the same activity with another person and see if you get the same result.

TEACHER INFORMATION

When we say something is hot or cold, we are usually thinking in terms of its comparison to something else. Boiling water isn't very hot compared with molten steel, for instance. We sometimes get a glass of water from the cold water tap and complain about the water being warm, but if we were to take a shower in water of the same temperature we would probably consider it very cold. This activity is designed to point out that our reaction to temperatures are relative. We frequently judge something to be hot or cold relative to the temperature of the skin. The lukewarm water used in this activity should be such that it feels neither hot nor cold to the skin. When the blindfolded person puts one hand into it after the other hand is already in hot water, however, the lukewarm water will probably be described as cold or cool because of the contrast with the hot water. When one hand is already in cold water, the lukewarm water will feel warm.

ACTIVITY 132: How Many Touch Words Can You Find?

MATERIALS NEEDED

- Paper
- Pencils
- Bowl or small box

PROCEDURE

1. Do this activity with a group of your classmates.
2. Write a touch word (such as hard, soft, smooth, rough) on each of several small pieces of paper. The words used should describe objects that are in the room or otherwise accessible.
3. Fold the papers and put them in the bowl.
4. Divide the participants into two teams.
5. Draw a paper from the bowl and read the word.
6. Each team should find something that will match the word that is read. The first team to find an item to match the touch word gets a point.
7. Draw another paper. Repeat the process until the time is up or a predetermined number of points is reached.

TEACHER INFORMATION

This activity gives students an opportunity to use many descriptive words that are associated with the sense of touch and to match those words with objects they describe. Here is a beginning list of touch words you or your students can choose from in making the papers to go in the bowl:

Hard	Heavy	Smooth	Soft	Solid	Wet
Dry	Spongy	Hairy	Dusty	Stiff	Scratchy
Rough	Hot	Cold	Lukewarm	Slippery	Silky
Sticky	Bumpy	Lumpy	Bristly	Coarse	Woolly
Furry	Oily	Prickly	Limp	Gooey	Greasy
Gritty	Flabby	Fluffy	Fuzzy	Slimy	Squishy
Thorny	Stretchable	Springy	Slick	Cuddly	

As a follow-up activity, consider having the children think of some touch experiences that are sometimes associated with emotions or feelings. Have them describe touch experiences that make them feel warm, cold, itchy, tingly, frightened, excited, safe, happy, eerie, and so forth.

ACTIVITY 133: Where Is Touch Most Sensitive?

MATERIALS NEEDED

- Blindfold
- Variety of objects, such as a ball, a pencil, a book, an eraser, a piece of crumpled paper

PROCEDURE

1. Find a partner to help you with this activity.
2. Blindfold your partner.
3. Select one of the objects and place it on the table in front of your partner.
4. Tell your partner that he or she is to identify the object on the table by feeling it, but the hands are not to be used. The arms, face, feet, or anything but the hands may be used.
5. Have someone else try the same thing, but with a different object to identify.
6. Discuss the experience Was it easy? What part of the body did you want to use? What parts of the body have the greatest sense of touch?

TEACHER INFORMATION

The children should note that the sense of touch is the only one of the five senses that is not confined to the head area. The organ of the body that is usually associated with the sense of touch is the skin, although areas inside the body have varying amounts of touch sensitivity.

Not all areas of the skin have the same degree of sensitivity to touch. One of the areas of greater sensitivity is the fingertips. Because of this and the dexterity of the fingers, we usually use our fingers to feel something. This activity should help participants realize why we use our fingers to do this and how little sensitivity some of the other parts of the body have.

If you have been using a "feely box" for activities with the sense of touch, items from it would be excellent to use in the above activity. These are items the children have already experienced with the fingers and they will be able to compare the task of identifying them with other parts of the body.

Section 8

HEALTH AND NUTRITION

TO THE TEACHER

Every person needs to develop wise eating habits and understand some basic concepts and principles for proper care of the body. Some children have access to little food and need to be especially careful that what they eat is nutritious. Many have excessive amounts of junk food available to them and considerable social pressure or desire to consume unhealthful amounts of it. The activities of this section will help to increase awareness of good eating habits, proper sanitation, and care of the teeth.

Someone said commitment comes partly through ownership of an idea. This is a good time to get students involved in researching, writing, and sharing reports on topics related to the activities herein, and in teaching each other. The process of teaching someone else good health habits will increase the level of commitment on the part of the presenters to develop those habits themselves.

ACTIVITY 134: Why Do We Wash Our Hands Before Handling Food?

MATERIALS NEEDED

- Two small potatoes (washed)
- Potato peeler
- Two sterilized jars (with lids that will seal)
- Masking tape or two gummed labels
- Marker

PROCEDURE

1. Without washing your hands, peel a potato and put it in one of the jars.
2. Scrub your hands well, using soap, then peel the other potato and put it in the other jar. Be sure to wash the potato peeler first.
3. Seal both jars with their lids.
4. Using gummed labels or masking tape, label the first jar "Hands unwashed" and the second "Hands washed".
5. Place the two jars in a warm place where they can be observed but unmolested for several days.
6. Leaving the lids on the jars, examine the two potatoes daily. Compare them. Do you see any changes? If so, write a description of the changes you see.

TEACHER INFORMATION

The potato peeling should be done by someone whose hands appear to be clean but have not been washed for several hours. Be sure students notice that the hands appear to be clean. After a few days, mold is likely to form on the potato peeled with unwashed hands, as mold spores that were transferred from the hands to the potato multiply. Little or no growth will be noted on the potato that was peeled after the hands were scrubbed.

Daily observation, particularly after the mold begins rapid growth, should help students realize the importance of washing their hands before handling food. Remind students that the hands looked clean even before the first potato was peeled.

ACTIVITY 135: Why Are the Basic Food Groups Important?

MATERIALS NEEDED

- Paper and pencil
- Encyclopedia
- Health books
- Other books about food
- Small group of students

PROCEDURE

1. With your small group, select one of the following basic food groups to study:
   a. Vegetables and fruits
   b. Meat
   c. Bread and cereals
   d. Milk and milk products
2. Search the reference books you have for answers to the following questions, regarding your food group:
   a. What do these foods have in common?
   b. What do these foods do for our bodies?
   c. What vitamins and minerals do these foods provide?
   d. How many calories are usually in a normal serving?
   e. Are these foods expensive?
3. Report your findings to the class.

TEACHER INFORMATION

A balanced diet is necessary for our bodies to have energy to keep warm, to do our work and to play. Our bodies must also have materials for repair and growth and to keep the organs functioning properly.

This activity works well as a group activity, with each group of students assigned a different food group to explore. When they have gathered their information, have each group share their findings with the rest of the class. Preparation for these reports provides an excellent opportunity for students to practice their skills at illustrating with pictures and compiling information into charts. With all information in, have students plan a balanced diet, using the information learned in this activity. This could be done either by individuals or in groups, according to the interest and ability of the students.

ACTIVITY 136: What Are the Most Common and the Most Popular Foods?

MATERIALS NEEDED

- Copies of the "Food I Like and Food I Eat" activity sheet
- Chart paper
- Pencils
- Marker

PROCEDURE

1. Make several copies of the "Food I Like and Food I Eat" activity sheet and write "Breakfast" at the top of each.
2. Give a copy of this sheet to each of several people and ask them to write the food items they *most commonly eat* for breakfast (at least once each week) in the left column and the foods they *prefer* for breakfast in the right column.
3. Collect the sheets.
4. Write all the food items people included in the left column on a "Common and Popular Foods" chart.
5. For each item in the list, count the number of people who eat that item regularly and write the number in the "How Many Eat" column on the chart.
6. For each item in the list, count the number of people who prefer that item and write that number in the "How Many Prefer" column on the chart.
7. Display the chart on the wall.

TEACHER INFORMATION

This activity could be done by an individual or a committee. A class discussion of the popular and common foods and the reasons we eat what we do could be a valuable follow-up. The discussion should include the relative nutritional value.

While this activity is being conducted, other individuals or groups could be gathering similar information for lunch and dinner foods.

FOOD I LIKE AND FOOD I EAT

Meal: ______________________

Food I eat most:

Food I like most:

COMMON AND POPULAR FOODS

Food Item	How Many Eat	How Many Prefer

ACTIVITY 137: How Well Balanced Is Your Diet?

MATERIALS NEEDED

- Paper
- Pencils
- Encyclopedia or health text

PROCEDURE

1. Write a "Three-Meal Menu" of everything you eat for breakfast, lunch, and dinner today. Do this for three days.
2. Find a list of the four basic food groups in your encyclopedia or health book.
3. Examine your menus of food you consumed for three days. Beside each food item, write which food group that item belongs to.
4. At the bottom of each "Three-Meal Menu," write the names of the four food groups and the number of items you consumed that day from each group.
5. Do you show a balanced diet (at least one serving from each of the four food groups) for each day? If not, write down which food group(s) you need to include in your diet more frequently.

TEACHER INFORMATION

If school lunch is served at your school, be sure to include it as one of the meals used for this activity.

Students should easily be able to find, in their health book or encyclopedia, reference to the four basic food groups (fruits and vegetables, breads and cereals, meat, and dairy products). If students have already completed Activity 136 and still have their record of food eaten, they could use that information instead of rewriting the food items for nine meals as indicated above.

After students have completed this activity, discuss their findings. They should be cautioned to include frequent servings of each of the four basic food groups in their diet (at least one from each group every day), and to avoid excessive amounts of sugar and of foods that contain a lot of fat, such as ice cream and fatty meat. You might even make a class chart (without names) showing the frequency of consumption of the four food groups. This will spotlight the extent to which the class is getting a balanced diet and which, if any, food groups are being short-changed.

ACTIVITY 138: How Can We Test Foods for Protein?

(Teacher-supervised activity)

MATERIALS NEEDED

- Lime powder
- Copper sulfate
- Water
- Two stirring sticks
- Two medicine droppers
- Small measuring cup
- Small amounts of common foods, such as meat, flour, butter, eggs, cheese, bread, salt, and sugar)
- Paper towels
- Two cup-size containers

PROCEDURE

1. Measure about two tablespoons of water into each of two containers.
2. Mix lime powder in one of the containers and copper sulfate in the other. Put in as much as you can get to dissolve.
3. Keep the two solutions separate. Put a clean dropper in each.
4. Put a small amount of meat on a paper towel.
5. Put two or three drops of lime water on one spot on the meat. Then put an equal amount of the copper sulfate solution on the same spot.
6. What happened?
7. Try putting equal amounts of the two solutions on some sugar.
8. Do the same with a piece of bread.
9. The different reactions you see are an indication of the amount of protein. Which of these substances has the most protein? Which does not have any?
10. Test the other foods you have and make a list, beginning with the foods having highest protein content and ending with those that have none.

TEACHER INFORMATION

Proteins help to make up the protoplast, or living portion, of body cells. Our bodies cannot manufacture their own protein, so we depend on the plants and animals we eat for our protein supply. Proteins are a vital food element, so it is important that we know which foods are the best sources of protein.

If a food containing protein is mixed with lime and copper sulfate, the mixture will turn to a violet color. The higher the protein content of the food, the darker the violet color will be.

ACTIVITY 139: Which Foods Contain Sugar?

MATERIALS NEEDED

- Empty food containers with labels intact
- Pencils
- Paper
- Chart paper
- Markers

PROCEDURE

1. For the food containers you have, write a list of the foods. Put a check mark beside each one that you think contains sugar.
2. Select one of the food containers. Write the name of the food on a chart.
3. Look at the list of contents on the container. Does the list include sugar? If so, write the type(s) of sugar used. Here are some types of sugar that might be included:

   Sugar
   Corn syrup
   Maltose
   Sucrose
   Fructose
   Corn sweetener
   Syrup
   Dextrose
   Glucose
   Lactose
   Molasses

4. The food contents are listed in order of quantity used. If one or more types of sugar are in the list, indicate on the chart whether they are listed first, second, third, and so forth.
5. Continue steps 2 to 4 for the food containers you have.
6. Now go back to the list you made in step 1 and put an X beside those foods you found do contain sugar. Compare these with your predictions (check marks from step 1).
7. Are you surprised?

TEACHER INFORMATION

The object of this activity is to make students aware that most foods we eat contain sugar and that sugar comes in many different forms. It is important that students complete step 1, complete with their predictions. Their findings will be more meaningful as they compare them with their own predictions.

ACTIVITY 140: Which Foods Contain Starch?

MATERIALS NEEDED

- Tincture of iodine
- Stirring sticks
- Eyedropper
- Paper cups
- Cornstarch
- Raw potato
- Bread
- Other foods, such as cooked egg white, cooked macaroni, meat, sugar, salt, crackers, and boiled rice
- Paper and pencils

PROCEDURE

1. Fill a paper cup about half full of water and add a small amount of cornstarch.
2. Using the dropper, add a drop or two of iodine to the cornstarch solution and stir. The blue-black color indicates the presence of starch.
3. Using the dropper, place a drop of tincture of iodine on a slice of raw potato.
4. What happened? Does potato contain starch?
5. Use the iodine test on a variety of other foods.
6. Make a list of those foods that contain starch and a list of those that do not.

TEACHER INFORMATION

Have one or more students look up starch in the encyclopedia or other reference books and prepare a short report on the nutritional value of starchy foods. They will learn that carbohydrates, including sugar and starch, provide heat and energy and are a necessary part of the diet. Excessive amounts, however, can be harmful.

ACTIVITY 141: Which Foods Contain Fat?

MATERIALS NEEDED

- Water
- Eyedropper
- Paper
- Pencils
- Brown paper bag
- Butter (melted)
- Variety of foods, such as salad dressing, boiled egg, leafy vegetable, meat, bread, nuts

PROCEDURE

1. Tear the brown paper bag up into small pieces, about 5 cm. (2 in.) in diameter.
2. Put a small amount of butter on one of the small pieces of paper. If the butter isn't melted, hold the paper in the sun for a few minutes.
3. Put a few drops of water on a second piece of paper.
4. Hold both pieces of paper up to the sunlight. Notice they are both translucent, letting some light through.
5. Leave both papers in the sun until the water dries.
6. Hold both pieces of paper up to the sunlight again. Are they both still translucent?
7. Fatty foods, like the butter, leave a permanent stain on the paper. Use the brown paper test on several other foods.
8. Make a list of foods you found to contain fat and those in which you did not detect fat.

TEACHER INFORMATION

Have one or more students look up fatty foods in their health book or the encyclopedia and report to the class. They will learn that fats provide energy. If excessive amounts of carbohydrates are consumed, these are changed to fat and stored in the body. Fats have a much higher calorie content than do carbohydrates.

ACTIVITY 142: How Do People Count Calories?

MATERIALS NEEDED

- Encyclopedia or cookbook
- Calorie chart for each student

PROCEDURE

1. For one full day, record on a "Calories for a Day" chart everything you eat. Be sure to include the number of servings of each item.
2. Find a calorie chart in the encyclopedia or cookbook.
3. Determine as closely as you can the number of calories in each of the food items you consumed. Record these numbers with the foods on your list.
4. Compare the number of calories you consumed with the average indicated for your age and height.
5. Make a realistic judgment as to whether the number of calories you consumed is appropriate. Don't be concerned about your total calories being a little above or below average. You might want to ask your teacher or your parent for his or her opinion.
6. If you determine that you should change your calorie intake, make adjustments in your diet and keep track of what you eat for an additional day or more.
7. Compare the results from step 6 with those from the first day. How did you do?

TEACHER INFORMATION

Caution students that to some people the number of calories eaten is considered personal information, so they shouldn't ask others for information from their lists. Before, during, or after this activity is an excellent time to invite a school nurse (or other available nurse, doctor, or nutrition expert) to talk to the group about nutrition and normal calorie intake.

ACTIVITY 143: How Many Calories Do You Use?

MATERIALS NEEDED

- Three copies of "My Daily Exercise Chart" for each student
- Calorie table
- Pencils

PROCEDURE

1. Keep a record of the types of exercise you do for one full day and the number of minutes you spend with each one. Use "My Daily Exercise Chart."
2. From information in the calorie table, write on your chart the number of calories used per minute for each exercise you listed.
3. Compute the total number of calories you used in each exercise and record it on your chart.
4. Determine the total number of calories you used in exercising during the entire day.
5. Keep a record of your exercising for an additional two or three days and compare with the first day. What kinds of exercises are you doing most? How many calories do you seem to be using in exercising on an average day?

TEACHER INFORMATION

A calorie table can be found in a health book or encyclopedia. In addition to keeping track of their own calories used in exercising, students might enjoy comparing their charts with those of others. Make a class chart, without names, showing the types of exercises that seem to be the most popular and the number of calories used by each. Total numbers of calories used daily in exercising can also be charted without names so each student can compare his or her results with those of the group. Students should understand that these figures represent only part of the total calories used. The body uses calories constantly, whether it is walking, running, resting, or even sleeping.

Name ____________________ Date ____________________

MY DAILY EXERCISE CHART

Type of Exercise	No. of Minutes Spent Doing It	No. of Calories Used per Minute	Total No. of Calories Used During This Exercise

Total No. of Calories Used Exercising Today

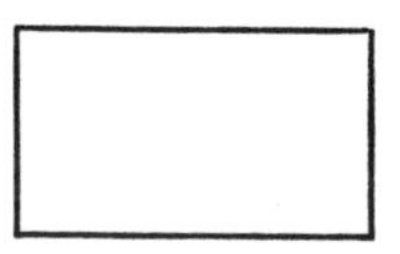

ACTIVITY 144: How Are Teeth Shaped for Their Task?

MATERIALS NEEDED

- Model of a full set of teeth
- Encyclopedia and health books
- Pencil and paper
- Mirror

PROCEDURE

1. Examine the model set of teeth and notice the different shapes.
2. Which teeth are sharpest? Which are most pointed? Which are flattest?
3. Use the mirror to look at your own teeth. Are yours shaped about the same as those in the model?
4. How are the front teeth shaped?What do you think they do best: cut, tear, or crush? How many of these are there? Write your answers on a sheet of paper.
5. How many teeth come to a single point? What do you think they do best? Write your answers on the paper.
6. Complete the paper by answering the same questions for teeth with two points and for those that are broad and flat.
7. When you are through, look up "Teeth" in the encyclopedia and find the names of the four types of teeth on your paper. Read about them and see if the information you wrote agrees with that in the encyclopedia.

TEACHER INFORMATION

A full set of adult teeth includes the following types and numbers in the upper jaw (the same types and numbers are found in the lower jaw):

Four *incisors,* located in the center and front of the mouth. These are sharp and are used for cutting food.

Two *cuspids,* located at the corners of the mouth. These are pointed and are used to tear food.

Four *bicuspids,* located just behind the cuspids. These are used to tear and crush food.

Six *molars,* located at the back of the mouth. These are broad and flat and are used for grinding food.

In addition to examining human teeth and considering the specialized jobs they perform, students might enjoy comparing these with the teeth of various animals. Note the differences in the teeth of carnivorous and herbivorous animals.

This is an excellent time to invite a dental technician into your classroom to talk about teeth and their proper care.

ACTIVITY 145: What Does Tooth Decay Look Like?

MATERIALS NEEDED

- A healthy human tooth (or model)
- A decayed human tooth (or model)
- X-ray showing tooth decay
- Mirror
- Paper
- Pencil or crayons

1. Examine the samples of healthy and decayed teeth. What differences can you see?
2. Use the mirror to look in your own mouth. Which of your teeth are most like the samples?
3. While you are examining your own teeth, look for places that are discolored or for possible signs of tooth decay.
4. Examine the X-ray. Can you tell where there is tooth decay? If not, ask your teacher to help you.
6. Draw a picture of a healthy tooth and a decayed tooth. Then draw a picture of what you think each of them would look like in an X-ray. You might need to look at the sample X-ray again and notice where the light and dark colorings are.

TEACHER INFORMATION

If sample teeth or models cannot be obtained at your dentist's office, ask your school nurse to help you locate some samples. If they are not available, use pictures, but the real thing should leave a more lasting impression on students. You should be able to obtain one or more X-rays from your dentist. When you get the sample X-ray, have the dentist, a technician, the school nurse, or other qualified person brief you on reading the X-ray.

During discussions related to this activity, point out that the outer layer of the tooth is a hard enamel covering. When decay extends through this protective shell, the decay progresses more rapidly and becomes painful. This is one reason it is important for such problems to be detected and corrected early.

ACTIVITY 146: How Can You Clean Your Teeth Best?

MATERIALS NEEDED

- Plaque-indicator pills (one for each student)
- Toothbrushes (one for each student)
- Toothpaste
- Toothpicks
- Dental floss
- Model set of teeth
- Sink and water
- Paper cups

PROCEDURE

1. Chew up the pill.
2. Rinse your mouth with water and notice how much of the red coloring is left on your teeth.
3. Brush your teeth using a back-and-forth motion. Rinse your mouth and notice how much of the red coloring is left in your mouth.
4. Brush your teeth again, this time using an upward motion on your lower teeth and a downward motion on the upper teeth. Rinse your mouth and notice how much of the red coloring is left in your mouth.
5. Carefully use a toothpick to clean places where the toothbrush did not clean. Rinse your mouth and check for red coloring again.
6. Use dental floss between your teeth. Rinse your mouth and check for red coloring again.
7. Did each of these methods clean out some of the red coloring that was missed by the others? What can you say about that?

TEACHER INFORMATION

If possible have the students bring their toothbrushes from home wrapped in plastic or foil. Each of the techniques used for cleaning teeth in this activity should clean out some of the red coloring that was missed by the other methods. Relative effectiveness in cleaning food out of the teeth is similar. Food that is between the teeth and missed by the toothbrush continues to provide feeding ground for bacteria. The waste product of the bacteria is called plaque. Plaque contains acid, which decays the teeth.

ACTIVITY 147: How Fast Are Your Reactions?

MATERIALS NEEDED

- Meter stick

PROCEDURE

1. In this activity, you will test your reaction time. Reaction time is one indicator of health condition.
2. Have your partner hold the meter stick vertically. Your partner should hold it at the top and the lower end should be between your thumb and index finger.
3. Ask your partner to drop the meter stick without warning. When the stick drops, grasp it as quickly as you can with your thumb and index finger. Note how far it fell by reading the centimeter scale where you grasped the meter stick.
4. Try it three or four times and see if you can improve your reaction time.
5. Trade places with your partner. This time you drop the meter stick.
6. Practice and see if you can both improve your reaction time.
7. Have you ever had a physical examination by a doctor? If so, did the doctor check the reflexes of your knees?

TEACHER INFORMATION

Students will enjoy comparing reaction times and trying to improve their own with this and other reaction-testing activities you might care to use. For example, have each student place a dime on the back of his or her hand, then tip it off and try to catch it before it hits the floors. Try left hand, right hand, and both at the same time. A competitive reaction-time activity is the hand slapper. The first person holds a hand palm up and the second places his or her hand on that of the first, palm down. Person 1 tries to slap the back of the hand of person 2 before person 2 can move out of the way. Again, try each hand separately and both together. Caution students not to slap hard enough to hurt their partners.

Bibliography

Selected Professional Texts

Blough, Glenn O., and Julius Schwartz. *Elementary School Science and How to Teach It* (7th ed.). New York: Holt, Rinehart & Winston, 1984.

Carin, Arthur, and Robert B. Sund. *Teaching Science Through Discovery* (5th ed.). Columbus, OH: Charles E.Merrill, 1985.

Esler, William K. and Mary K. Esler. *Teaching Elementary Science* (4th ed.). Belmont, CA: Wadsworth 1984.

Gega, Peter C. *Science in Elementary Education* (4th ed.). New York: John Wiley & Sons, Inc., 1982.

Jacobsen, Willard J., and Abby Barry Bergman. *Science for Children.* Englewood Cliffs, NJ: Prentice Hall, 1980.

Kauchak, Donald, and Paul Eggen. *Exploring Science in Elementary Schools.* Chicago: Rand McNally, 1980.

Rowe, Mary Budd. *Teaching Science as Continuous Inquiry: A Basic* (2nd ed.). New York: McGraw-Hill, 1978.

Victor, Edward. *Science for the Elementary School* (5th ed.). New York: MacMillan Publishing Company, 1985.

Periodicals

Astronomy. AstroMedia Corp., 625 E. St. Paul Ave., Milwaukee, WI 53202
Audubon. National Audubon Society. 950 Third Ave., New York, NY 10022
Discover. Time Inc. 3435 Wilshire Blvd., Los Angeles, CA 90010
National Geographic. National Geographic Society, 17th and M Sts. N.W., Washington, DC 20036
Natural History. American Museum of Natural History, Central Park West at 79th St., New York, NY 10024
Ranger Rick's Nature Magazine. National Wildlife Federation, 1412 16th St. N.W., Washington, DC 20036
Science. American Association for the Advancement of Science, 1515 Massachusetts Ave. N.W., Washington, DC 20005
Science and Children. National Science Teachers Association, 1742 Connecticut Ave. N.W., Washington, DC 20009
Smithsonian. Smithsonian Associates, 900 Jefferson Dr., Washington, DC 20560
**World.* National Geographic Society, 17th and M Sts., Washington, DC 20036
**Zoo Books.* Wildlife Education Ltd., 930 West Washington St., San Diego, CA 92103

*For Elementary Age Students

Note: Check on subscription addresses. Many periodicals are now using the National Data Center in Boulder, Colorado rather than their national headquarters.

Sources of Free and Inexpensive Materials for the Life Sciences

Before requesting free and inexpensive materials, consider the following:

1. Use school stationery whenever possible. Most suppliers prefer it; some require it.
2. When you are requesting free materials, it is an act of courtesy to include a self-addressed, stamped envelope.
3. Do not ask for excessive amounts of free materials. Remember, the suppliers are generously paying the costs.
4. Be specific in your requests.
5. A word of thanks is in order at the time of your request and upon receipt and use of the materials.

A.I. Root Company
Medina, OH 44256
(pamphlets about bees)

American Heart Association
c/o your local chapter
(heart drawings; pamphlets and handouts about the heart and how it works; catalog of films about the heart)

American Indian Archaelogical Institute
Route 199, P.O. Box 260
Washington, CT 06793
(pamphlets about ancient Indian sites and common tools used)

Animal Welfare Institute
P.O. Box 3650
Washington, DC 20007
(pamphlets about attitudes toward animals; catalog of information with prices)

Chemical Manufacturers Association
2501 M Street, N.W.
Washington, DC 20037
(booklet about food additives; pamphlet entitled "Chemicals in Normal Everyday Foods"; a guide and glossary entitled "Consumers Chemistry")

U.S. Forest Service
12th and Independence, S.W.
P.O. Box 2417
Washington, DC 20013
(pamphlets entitled "Suggestions for Incorporating Forestry into the School Curriculum" and "Investigating Your Environment")

Garden Club of America
598 Madison Avenue
New York, NY 10022
(environmental education packet entitled "The World Around You" that includes one free packet per teacher, with additional copies available for a small charge)

The Hogg Foundation
University of Texas
Austin, TX 78710
(price list of films and booklets about mental health)

Kellogg's
Battle Creek, MI 49016
(catalog of food and nutrition publications, limited copies free)

Massachusetts Society for the Prevention of Cruelty to Animals (MSPCA)
450 Salem End Road
P.O. Box 2314
Framingham Centre, MA 01701
(catalog of booklets, teaching aids, and posters about animal care and abuse, prices included)

National Audubon Society
1130 Fifth Avenue
New York, NY 10001
(catalog of materials dealing with birds, wildlife, energy, etc., prices included)

National Coal Association
Coal Building
1130 17th Street, N.W.
Washington, DC 20036
(publications about coal, including an activity program for grades K-3 entitled "Discovering Coal")

National Science Teachers Association
1742 Connecticut Avenue, N.W.
Washington, DC 20009
(catalog of science publications, posters, and other learning aids, prices included)

Sunkist Growers, Inc.
Consumer Services
P.O. Box 7888
Van Nuys, CA 91409
(booklets entitled "Build a Better You with Fresh Citrus" and "Questions and Answers About Vitamin C and Fresh Citrus Fruit" available in classroom quantities; booklet entitled "Sunkist Fresh Citrus Handbook" available in single copies; series of posters entitled "Think Orange When You Choose a Snack" and "Vitamin C Is Not Stored in the Body So You Need It Every Day" available in single copies)

Science Supply Houses

Carolina Biological Supply Co.
2700 York Rd.
Burlington, NC 27215

Central Scientific Company
2600 South Kostner Avenue
Chicago, IL 60623

Edmund Scientific
101 E. Gloucester Pike
Barrington, NJ 08007
(Catalog for industry and education)

Fisher Scientific Company
4901 West Lemoyne
Chicago, IL 60651

Flight Systems, Inc.
9300 East 68th Street
Raytown, MO 64133
(Model rocketry)

Frey Scientific
905 Hickory Lane
Mansfield, OH 44905
(General science catalog)

Markson Science, Inc.
7815 S. 46th St.
Phoenix, AZ 85040
(Similar to Edmund)

MMI Space Science Corp.
2950 Wyman Parkway
P.O. Box 19907
Baltimore, MD 21211
(Astronomy and space science, teaching materials reference catalog)

Sargent-Welch Scientific Co.
7300 N. Linder Ave.
P.O. Box 1026
Skokie, IL 60077
(General science supply, similar to Frey)

Ward's Natural Science Establishment
3000 Ridge Road East
Rochester, NY 14622